R.W. Hasse   W.D. Myers

Geometrical Relationships of Macroscopic Nuclear Physics

With 33 Figures

Springer-Verlag Berlin Heidelberg New York
London Paris Tokyo

Dr. Rainer W. Hasse
Gesellschaft für Schwerionenforschung mbH, D–6100 Darmstadt and
Kernforschungszentrum Karlsruhe, D-7500 Karlsruhe, Fed. Rep. of Germany

Dr. William D. Myers
Gesellschaft für Schwerionenforschung mbH, D–6100 Darmstadt, Fed. Rep. of Germany
and Lawrence Berkeley Laboratory, University of California, Berkeley, CA 94720, USA

ISBN-13:978-3-642-83019-8   e-ISBN-13:978-3-642-83017-4
DOI: 10.1007/978-3-642-83017-4

© Springer-Verlag Berlin Heidelberg 1988
Softcover reprint of the hardcover 1st edition 1988

2153/3150 – 543210

Preface

The aim of this book is to provide a single reference source for the wealth of geometrical formulae and relationships that have proven useful in the description of atomic nuclei and nuclear processes. While many of the sections may be useful to students and instructors it is not a text book but rather a reference book for experimentalists and theoreticians working in this field. In addition the authors have avoided critical assessment of the material presented except, of course, by variations in emphasis.

The whole field of macroscopic (or Liquid Drop Model) nuclear physics has its origins in such early works as [Weizsäcker 35] and [Bohr 39]. It continued to grow because of its success in explaining collective nuclear excitations [Bohr 52] and fission (see the series of papers culminating in [Cohen 62]). These developments correspond to the first maximum in the histogram below, showing the distribution by year of the articles cited in our Bibliography. After the Liquid Drop Model had been worked out in some detail the development of the Strutinsky approach [Strutinsky 68] (which associates single particle contributions to the binding energy with the shape of the nucleus) gave new life to the field. The growth of interest in heavy-ion reaction studies has also contributed. Currently the rate at which new developments are occurring is decreasing and this provides us with the opportunity to bring the material together in a meaningful

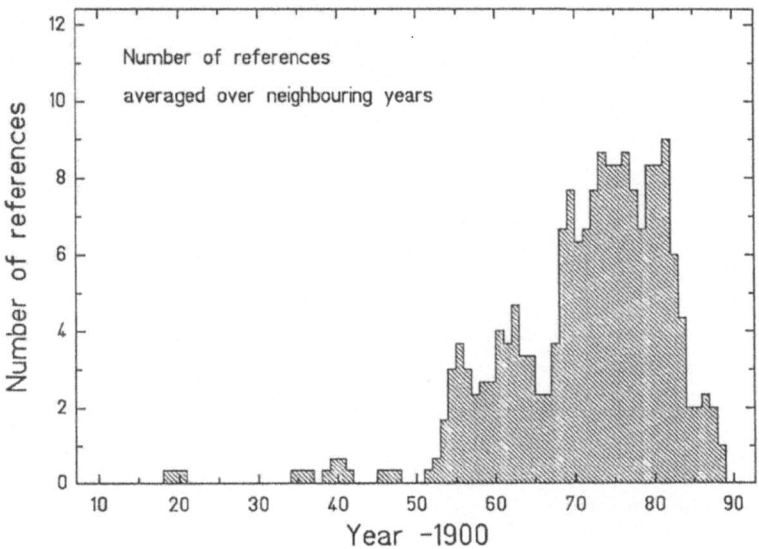

way. Some material is also included from the fields of medium- and high-energy nuclear collisions and this section can be expected to grow in future editions.

The idea for this book was born in the middle 1970's which corresponds to the location of the broad peak in the histogram below. Its realization was delayed until the invention of the computer typesetting program $\TeX$* and the higher level document preparation system $\LaTeX$. These tools made it possible for the authors to complete the preparation of a photo ready manuscript in their spare time over a period of nine months after a decade of carrying around boxes of source material.

In the Index and in the list of Citations we have referenced the immediately following equation number rather than page numbers. This nonstandard approach was chosen because it is more precise and because it could be done automatically.

There are many ways for errors to find their way into a compilation such as this one. Indeed, the original sources were not completely error free. We've done our best to be accurate and will endeavor to supply errata to anyone who requests it and supplies us with their name and address. If there is enough new material we'll consider sending out an update from time to time until a new edition is warranted.

The authors are indebted to their colleagues for bringing useful material to their attention and for checking various parts of the manuscript. They were inspired in their work by association with Wladek Swiatecki, who has made many contributions to the field, and by their interactions with Hans Krappe, Peter Möller, Ray Nix, Adam Sobiczewski, Vitaly Pashkevich and many others. The basic approach we have employed was suggested by Nikolai Ivanovich Lobachevski [†].

Finally, we want to express our gratitude to the management of GSI (Gesellschaft für Schwerionenforschung Darmstadt), especially Peter Armbruster and Wolfgang Nörenberg for their enthusiastic endorsement of this undertaking. In particular, we like to acknowledge the technical support of Enrico Pfeng who actually entered most of the material into $\LaTeX$.

RAINER W. HASSE

WILLIAM D. MYERS

---

[*] $\TeX$ is a trademark of the American Mathematical Society.

[†] As quoted by Tom Lehrer.

Contents

Chapter 1

Definitions and Notation

1.1 Introduction

A nucleus is characterized by its proton and neutron numbers Z and N. Their sum, the mass number $A = N + Z$, is also frequently used. One of the reasons for A being an important quantity is that the volume of a nucleus is, to a very good approximation, simply proportional to the number of particles. This property is associated with the saturation of nuclear forces and it permits the formulation of a macroscopic approach to the description of certain aspects of the nuclear many-body problem. This approach focuses its attention on the degrees of freedom describing the shape of the nuclear surface which, although not perfectly sharp, is known experimentally to be fairly well defined, except for small nuclei.

The reasons behind the applicability of such a macroscopic (or Liquid Drop Model, LDM) approach have nothing to do with any similarity between the interactions of nucleons and the classical, short mean-free path interactions of atoms in a water droplet. The crucial requirement is that deviations from bulk behavior should be confined to a (relatively thin) surface layer, a condition that can hold even in the case of quantized, weakly interacting (or even noninteracting) particles. These considerations are carefully developed in [Błocki 77], reviewed in [Myers 82] and are, of course, usually discussed in any major survey work on nuclear physics such as [Bohr 69], [Eisenberg 70] or [Ring 80].

1.1.1 Liquid Drop Model

Originally conceived more than 50 years ago ([Weizsäcker 35]) for the purpose of calculating ground state nuclear binding energies, the Liquid Drop Model began to assume a wider range of applicability when it was recognized that the gross properties of nuclear fission could be understood in terms of the shape dependence of the surface and electrostatic energies. In its simplest formulation the Liquid Drop Model mass formula can be written

$$M_{\mathrm{LD}}(N, Z) = M_N N + M_Z Z - a_{\mathrm{V}} A + a_{\mathrm{S}} A^{2/3} B_{\mathrm{surf}} + a_{\mathrm{C}} \frac{Z^2}{A^{1/3}} B_{\mathrm{Coul}} , \qquad (1.1)$$

where the assumption has been made that the nuclear matter in the interior is uniform and incompressible so that the radius of a spherical nucleus is proportional to $A^{1/3}$.

The quantities M_N and M_Z in (1.1) are the masses of the individual neutrons and protons and the nuclear mass M_{LD} is just the sum of these individual masses reduced by the nuclear binding energy which is given to a good approximation by the last three terms. The first of these terms called the *volume energy* is proportional to the total number of particles. The coefficient a_V is usually written as

$$a_V = a_1(1 - \kappa_V I^2) \,, \tag{1.2}$$

where a_1 is the binding energy per particle of symmetric (i.e. $N = Z$) nuclear matter and the term proportional to I^2 (where $I = (N - Z)/A$) is included in order to describe the dependence of the bulk binding energy on the neutron excess. Sometimes additional terms are included in (1.2) in order to treat the finite compressibility of the nuclear medium and its dependence on the neutron excess [Myers 74]. In its simplest form (1.2) has no shape dependence. The next term in (1.1) is the *surface energy* whose coefficient can be written

$$a_S = a_2(1 - \kappa_S I^2) \,, \tag{1.3}$$

in analogy with (1.2). The factor $A^{2/3}$ is proportional to the surface area for a spherical nucleus and the quantity B_{surf} relates the surface energy of a deformed nucleus to that of a spherical nucleus with the same volume. Many extensions of this simple formulation are possible. One of the most significant is discussed below in Sect. 1.3.2 in connection with the finite range force surface energy. The last term in (1.1) is the *Coulomb energy*. For a spherical nucleus $Z^2/A^{1/3}$ is proportional to Q^2/R (the ratio of the total charge squared to the radius). The final factor in this term, B_{Coul}, relates the actual Coulomb energy of a deformed nucleus to that of a spherical nucleus of the same volume.

The subject matter of this book consists, in part, of various mathematical descriptions of nuclear shapes and the calculation of the corresponding B_{surf} and B_{Coul}. Extensions of (1.1) involve the inclusion of finite compressibility effects, diffuseness corrections to the Coulomb energy and other refinements that often necessitate the definition of additional quantities B_i. Some of these developments are reviewed in [Myers 82]. The proceedings of recent conferences on nuclear masses such as [Klepper 84] should be consulted for more details.

1.1.2 Droplet Model

One specific approach to extending the LDM is called the Droplet Model. This line of development involves the formulation (in [Myers 69] and [Myers 74]) of an expression that includes all the LDM terms and in addition all the terms that arise when the macroscopic approach is extended to one higher order in the expansion parameters $A^{-1/3}$ and I^2. The extensions are based on removing the assumption of incompressibility and removing the requirement that the neutron and proton density distributions have a common surface. These generalizations result in a mass formula with more terms than the LDM and one which also predicts changes in nuclear radii which are in better agreement with the measured values. Since the energy associated with the volume redistribution of charge under the influence of the Coulomb forces and related surface energy terms are included in the model, the binding energy expression is somewhat complicated. After minimization with respect to the various degrees of freedom contained in the model an expression results having the form

$$
\begin{aligned}
E(N, Z; \text{shape}) \; = \; & \left[-a_1 + J\bar{\delta}^2 - \frac{1}{2}K\bar{\epsilon}^2 + \frac{1}{2}M\bar{\delta}^4 \right] A \\
& + \left[a_2 + \frac{9}{4}\frac{J^2}{Q}\bar{\delta}^2 \right] A^{2/3} B_{\text{surf}} + a_3 A^{1/3} B_{\text{curv}} \\
& + c_1 Z^2 A^{-1/3} B_{\text{Coul}} - c_2 Z^2 A^{1/3} B_{\text{red}} \\
& - c_5 Z^2 B_{\text{w}} - c_3 Z^2 A^{-1} - c_4 Z^{4/3} A^{-1/3} \; ,
\end{aligned} \tag{1.4}
$$

where

$$
\bar{\delta} \; = \; \frac{I + \frac{3}{16}\frac{c_1}{Q}ZA^{-2/3}B_{\text{v}}}{1 + \frac{9}{4}\frac{J}{Q}A^{-1/3}B_{\text{surf}}} \; ,
$$

$$
\bar{\epsilon} \; = \; [-2a_2 A^{-1/3}B_{\text{surf}} + L\bar{\delta}^2 + c_1 Z^2 A^{-4/3}B_{\text{Coul}}]/K \; , \tag{1.5}
$$

and the constants $c_1 \cdots c_5$ are related by the expressions

$$
c_1 \; = \; \frac{3}{5}\frac{e^2}{r_0}
$$

$$
c_2 \; = \; \frac{c_1^2}{336}\left(\frac{1}{J} + \frac{18}{K} \right)
$$

$$
c_3 \; = \; \frac{5}{2}c_1 \left(\frac{b}{r_0} \right)^2
$$

$$
c_4 \; = \; \frac{5}{4}c_1 \left(\frac{3}{2\pi} \right)^{2/3}
$$

$$
c_5 \; = \; \frac{1}{64}\frac{c_1^2}{Q} \; . \tag{1.6}
$$

The shape dependent functions B_{surf} and B_{Coul} are already familiar from our discussion of the LDM in the previous section. The new quantities B_{curv}, B_{red}, B_{v}, and B_{w} concern: (1) The curvature correction to the surface energy; (2) The redistribution of charge in the interior of the nucleus arising from the Coulomb forces and (3) Two terms similar to B_{red} but having to do with Coulomb redistribution effects in the surface. These quantities are defined below in Sect. 1.6. The coefficients occurring in (1.4) are a_1, J, a_2, Q and r_0 which are analogous to the LDM coefficients $a_1, \kappa_{\text{V}}, a_2, \kappa_{\text{S}}$ and a_{C} of (1.1). In addition there are three new coefficients: K, the compressibility; L, which concerns the dependence of compressibility on neutron excess, and M, which is the coefficient of a higher order term in $I = (N - Z)/A$.

1.2 Nuclear Radius Constant

The volume of a spherical nucleus is given by

$$V = \frac{4}{3}\pi R_0^3 \ , \tag{1.7}$$

and the relatively large incompressibility of nuclear matter leads to the possibility of writing

$$R_0 = aA^{1/3} \ , \tag{1.8}$$

where a is nearly constant for all nuclei. That this coefficient is not exactly constant is discussed in various places such as [Elton 61], [Myers 69], [Friedrich 82] and [Myers 83]. In earlier work that did not take possible variations in a into account it was possible to express the central density as

$$\varrho_0 = \left[\frac{4}{3}\pi r_0^3\right]^{-1} \ , \tag{1.9}$$

where r_0, often referred to as the *radius constant*, is a constant (whose value is ≈ 1.18 fm) and ϱ_0 is the *density of nuclear matter*. Then for a spherical nucleus containing A nucleons

$$R_0 = r_0 A^{1/3} \ . \tag{1.10}$$

Since many of the relationships given in this book were derived under the assumption (1.10) we will retain that perspective. In some cases specific generalizations of this approach will also be presented.

1.3 Geometrical Quantities

For an arbitrarily shaped uniform density distribution the expression (1.7) can be used to define the equivalent sharp spherical radius R_0. Dimensionless lengths can then be expressed in units of R_0 and various other quantities can be expressed in units corresponding to the value of the relevant quantity for a uniform sphere of radius R_0 and total mass M and charge Z.

1.3.1 Coordinate Systems

In Chaps. 6 and 7 the cylindrical coordinates* (P, z) are expressed in units of R_0 as (ϱ, ζ) and the spherical coordinate R is expressed in units of R_0 as r according to

$$
\begin{aligned}
P &= R_0 \varrho \ , \\
z &= R_0 \zeta \ , \\
R &= R_0 r \ .
\end{aligned} \tag{1.11}
$$

Shape functions in cylindrical coordinates * $P(z)$ or radial coordinates $R(\theta, \phi)$

---

*P stands for capital ϱ

can also be expressed in units of R_0 by the expressions

$$
\begin{aligned}
P(z) \equiv \varrho_s(z) &= R_0 \varrho(\zeta) , \\
R(\theta, \phi) &= R_0 r(\theta, \phi) .
\end{aligned}
\tag{1.12}
$$

Two other especially defined quantities characterizing specific shapes (half the length z_0 and the neck radius P_n) are found in Chaps. 6 and 7. In units of R_0 they can be written as ζ_0 and ϱ_n according to

$$
\begin{aligned}
z_0 &= R_0 \zeta_0 \\
P_n &= R_0 \varrho_n.
\end{aligned}
\tag{1.13}
$$

1.3.2 Radial Moments

The radial moments of a distribution $\varrho(\boldsymbol{r})$ are defined by the expression

$$
m_n = \int d^3 r \, r^n \varrho(\boldsymbol{r}) ,
\tag{1.14}
$$

where r is the radial distance from the center of mass and the function $\varrho(r)$ is assumed to be normalized so that $m_0 = 1$. For a uniform sphere of radius R_0,

$$
m_n = \frac{3}{n+3} R_0^n .
\tag{1.15}
$$

If (as in Chaps. 5 – 9) we define the mean square radius m_2 in units of R_0^2 by the quantity r_{rms}^2 according to the expression

$$
m_2 = R_0^2 r_{\mathrm{rms}}^2
\tag{1.16}
$$

then we can also write

$$
r_{\mathrm{rms}}^2 = \frac{3}{4\pi} \int d\tau \, r^2 \, \varrho(r) ,
\tag{1.17}
$$

where $d\tau = dV/R_0^3$ is the dimensionless volume element.

For a uniform sphere

$$
r_{\mathrm{rms}}^2 = \frac{3}{5} .
\tag{1.18}
$$

1.3.3 Multipole Moments

The general definition of the multipole moments of a distribution $\varrho(r)$ [Jackson 75] in terms of the spherical harmonics $Y_{\ell m}(\theta, \phi)$ can be written as

$$
M_{\ell m} = \int d^3 r \, r^\ell \, Y_{\ell m}(\theta, \phi) \, \varrho(\boldsymbol{r}) .
\tag{1.19}
$$

Axially symmetric distributions are conveniently characterized by the moments $\mathcal{Q}_\ell$, where

$$
\mathcal{Q}_\ell = \int d^3 r \, r^\ell \, P_\ell(\cos\theta) \, \varrho(\boldsymbol{r})
\tag{1.20}
$$

and

5

$$Q_\ell = \sqrt{\frac{4\pi}{2\ell+1}}\, M_{\ell 0}\,. \tag{1.21}$$

These moments Q_ℓ can be expressed in terms of the unit $3MR_0^\ell/8\pi$ and the relative multipole moments Q_ℓ of Chaps. 6 and 7 by the expression

$$Q_\ell = \frac{3}{4\pi} M R_0^\ell \, Q_\ell\,, \tag{1.22}$$

where M, the total mass, should be replaced by the total charge Ze for electric multipole moments. The dimensionless quantities Q_ℓ are defined by the expressions

$$
\begin{aligned}
Q \equiv Q_2 &= 2 \int \mathrm{d}\tau\, r^2\, \mathrm{P}_2(\cos\theta) \\
Q_\ell &= 2 \int \mathrm{d}\tau\, r^\ell\, \mathrm{P}_\ell(\cos\theta),
\end{aligned}
\tag{1.23}
$$

where $\mathrm{d}\tau = \mathrm{d}V/R_0^3$ is the dimensionless volume element and the radial distance r is also expressed in units of R_0.

1.3.4 Generalized Moments

In [Krappe 76]

$$\mathcal{M}_{\ell m} = 4\pi\, i^\ell \int \mathrm{d}^3 r\, j_\ell(kr)\, Y_{\ell m}(\theta,\phi)\, \varrho(\boldsymbol{r}) \tag{1.24}$$

and

$$\lim_{k\to 0} k^{-\ell} \mathcal{M}_{\ell m}(k) = 2\pi \left(\frac{i}{2}\right)^\ell \frac{\Gamma(1/2)}{\Gamma(\ell+3/2)}\, M_{\ell m}\,. \tag{1.25}$$

In [Davies 76a] the quantity $q_{\ell m}^{(k)}$ is defined by the expression

$$q_{\ell m}^{(k)} = \int \mathrm{d}^3 r\, r^\ell\, Y_{\ell m}(\theta,\phi)\, r^k \varrho(\boldsymbol{r}) \tag{1.26}$$

and folding model results are given for $k = 0, 2$. See also [Satchler 72] for $k = 0, 2, 4$.

1.3.5 Moments of Inertia

The rigid body moment of inertia for rotation about an arbitrary axis (taken to be the z axis in cartesian coordinates) can be written as

$$I_z = m \int \mathrm{d}^3 r\, (x^2 + y^2)\, \varrho(\boldsymbol{r})\,, \tag{1.27}$$

where m is the mass associated with a single particle and $\varrho(r)$ is the usual expression for the particle number density. Alternatively m can be replaced by M, the total mass of the object, and the density $\varrho(r)$ normalized to have a volume integral of unity.

If the symbol J is used to represent the moment of inertia relative to an axis passing through the center of mass then the moment of inertia of a uniform spherical distribution of radius R_0 is

$$J_0 = \frac{2}{5} M R_0^2 = \frac{(r_0/\text{fm})^2}{103.8415} \frac{\hbar^2 A^{5/3}}{\text{MeV}} . \tag{1.28}$$

For axially symmetric distributions dimensionless parallel and perpendicular moments of inertia are defined in units of the moment of inertia of a sphere (1.28) by the expressions

$$\begin{aligned}
\mathcal{J}_{\parallel} &= \frac{15}{8\pi} \int \mathrm{d}\tau \, \varrho^2 \\
\mathcal{J}_{\perp} &= \frac{15}{16\pi} \int \mathrm{d}\tau (r^2 + \zeta^2),
\end{aligned} \tag{1.29}$$

and the relative inverse effective moment of inertia reads

$$\mathcal{J}_{\text{eff}}^{-1} = \mathcal{J}_{\parallel}^{-1} - \mathcal{J}_{\perp}^{-1}. \tag{1.30}$$

1.3.6 Other Moments

As a measure of the deviation of the shape of a given uniform distribution from a sphere of equal volume the *relative mean squared radial distortion* [Myers 66] can be defined in units of R_0^2 by the expression

$$\overline{(\delta r)^2} = \int \frac{\mathrm{d}\Omega}{4\pi} \left(\frac{R(\theta)}{R_0} - 1 \right)^2 , \tag{1.31}$$

where $d\Omega$ is the solid angle element. It is also used by Carlson [Carlson 61b] to form a dimensioned quantity called the *anisotropy factor*

$$\langle r^2 \rangle = R_0^2 \left(1 + \overline{(\delta r)^2} \right) . \tag{1.32}$$

1.4 Surface Energies

Early in the development of nuclear physics [Weizsäcker 35] it was recognized that the binding energy of nuclei could be represented by a volume energy term proportional to the number of particles A which is reduced by surface energy and Coulomb energy contributions (see Sect. 1.1.1).

1.4.1 Geometrical Surface Energy

In general the surface energy of an arbitrarily shaped object can be written as

$$E_{\text{surf}} = \sigma \int \mathrm{d}S , \tag{1.33}$$

where σ is the surface tension (given, for example, in units of MeV/fm^2) and dS is simply the differential surface area element. For a sphere of radius R_0 the surface energy E_S is given by

$$E_S^0 = 4\pi R_0^2 \sigma . \tag{1.34}$$

Under the assumption that (1.10) holds, this expression can be rewritten as

$$E_S^0 = a_2 A^{2/3} \, , \tag{1.35}$$

where $a_2 = 4\pi r_0^2 \sigma$ is the usual Liquid Drop Model surface energy coefficient. For other shapes it is often convenient to express the relative surface energy B_{surf} in units of E_S^0, the surface energy of a sphere of equal volume

$$B_{\text{surf}} = \frac{E_{\text{surf}}}{E_S^0} = \frac{1}{4\pi} \int ds \, , \tag{1.36}$$

where ds is the surface element in units of R_0^2. The deformation part of the surface energy then reads

$$(B_{\text{surf}} - 1) \, E_S^0 \, . \tag{1.37}$$

1.4.2 Short Range Force Surface Energy

The nuclear part of the binding energy of a nucleus (when viewed as a lepto-dermous system, Chap. 2) can be expanded in a series consisting of a volume energy proportional to the number of particles, a surface energy and higher order energy terms of various kinds. An alternative approach is to simply fold (convolute) a phenomenological short range interaction into the density distri-bution using an expression like

$$E = \int d^3 r \int d^3 r' \, \varrho(\boldsymbol{r}) \, \varrho(\boldsymbol{r}') \, f(\boldsymbol{r} - \boldsymbol{r}') \, , \tag{1.38}$$

where ϱ is the assumed nuclear density distribution and f is the folding func-tion. Normally the folding function has two parameters (a strength and a range) and is only a function of $|\, \boldsymbol{r} - \boldsymbol{r}' \,|$. However, one could imagine more complex expressions that have different properties normal and transverse to the surface, or a strength that is a function of the local density and its derivatives (such as the Skyrme [Brack 85] and Gogny [Gogny 77] forces).

When the bulk binding energy (the total number of particles times the nuclear matter value for the binding energy per particle) is subtracted from Eq. (1.38) the resulting expression is sometimes [Krappe 73], [Krappe 79], [Möller 81] re-ferred to as the *Surface Energy*. Strictly speaking such an expression includes all the higher order terms beyond the leading order volume energy term that is explicitly removed and a more appropriate appellation might be *Surface Layer Energy* [Błocki 77].

One possible form for the folding function f in (1.38) is that of a finite square well [Preston 62]

$$f(r) = -V_P \, \Theta(r - a) \, . \tag{1.39}$$

For a uniform spherical density distribution of density ϱ_0, mass number A and radius R_0 the total interaction energy is

$$E_{\text{tot}} = -\frac{4}{3}\pi a^3 \varrho_0 V_P A \left[1 - \frac{9}{16} \left(\frac{a}{R_0} \right) + 0 + \frac{1}{32} \left(\frac{a}{R_0} \right)^3 \right] \, . \tag{1.40}$$

The first term in the square brackets is the volume energy term mentioned above (proportional to A), and the second is the surface energy (proportional to

$A^{2/3}$). There is no curvature correction (proportional to $A^{1/3}$) present here. Consequently all other representations of the nuclear interaction energy in terms of a simple momentum and density independent two-body force are also missing this term in the leptodermous expansion.

If a Yukawa interaction is employed and we restrict our discussion to uniform distributions of density ϱ_0 then (1.38) can be written

$$E_{KN} = -\frac{V_0}{4\pi a^3} \int d^3r \int d^3r' \frac{\exp(-\mid r - r' \mid /a)}{\mid r - r' \mid /a} . \tag{1.41}$$

The subscript KN in (1.41) denotes the work of Krappe and Nix [Krappe 73]. For a spherical distribution of radius R_0

$$E_{KN}^0 = 2\pi a V_0 R_0^2 \left[-\frac{2R_0}{3a} + 1 - \left(\frac{a}{R_0}\right)^2 + \left(1 + \frac{a}{R_0}\right)^2 \exp\left(-\frac{2R_0}{a}\right) \right] . \tag{1.42}$$

In (1.42), just as in (1.40), the term that would be proportional to $A^{1/3}$ is missing. In actual use [Krappe 73] the volume energy term, which is the first term in the square bracket of (1.42), is subtracted and the resulting expression, which contains the surface energy and the higher order terms, is referred to as a *generalized surface energy*.

Since the coefficient which multiplies the bracket in (1.42) is analogous to the usual expression for the surface energy of a sphere we can define a special quantity

$$\begin{aligned} E_{S,KN}^0 &= 2\pi a V_0 R_0^2 , \\ &= c_s A^{2/3} , \end{aligned} \tag{1.43}$$

where

$$c_s = 2\pi a V_0 r_0^2 . \tag{1.44}$$

Eq. (1.43) can then be identified with (1.35), the usual Liquid Drop Model expression for the surface energy. In analogy with (1.37) the deformation part of the Krappe-Nix energy associated with (1.41) can then be written as

$$(B_{KN} - 1) E_{S,KN}^0 , \tag{1.45}$$

where

$$B_{KN} = \frac{E_{KN} + E_{S,KN}^0 - E_{KN}^0}{E_{S,KN}^0} . \tag{1.46}$$

In exactly the same way as above the energy can be calculated using a Yukawa-plus-exponential form of interaction [Krappe 79] where the energy expression (1.38) becomes

$$E_{YE} = \frac{E_{S,YE}^0}{8\pi^2 a^4 R_0^2} \int d^3r \int d^3r' \left(2 - \frac{\mid r - r' \mid}{a}\right) \frac{\exp(-\mid r - r' \mid /a)}{\mid r - r' \mid /a} . \tag{1.47}$$

Here, as in (1.43) the quantity

$$E_{S,YE}^0 = c_s A^{2/3} \tag{1.48}$$

can be identified with the LDM expression for the surface energy (1.35). The two terms in the bracket in (1.47) have been chosen so that the volume energy term (the leading term in (1.42)) is missing. When this interaction is folded into the nuclear density distribution the leading term is the surface energy term. For a spherical distribution of uniform density ϱ_0 and radius R_0 the resulting interaction energy is [Krappe 79]

$$E_{YE}^0 = E_{S,YE}^0 \left[1 - 3 \left(\frac{a}{R_0} \right)^2 + \left(1 + \frac{R_0}{a} \right) \left(2 + 3 \frac{a}{R_0} + 3 \left(\frac{a}{R_0} \right)^2 \right) \exp \left(-\frac{2R_0}{a} \right) \right].$$
(1.49)

The deformation part of the Yukawa plus exponential energy corresponding to (1.47) can be written as

$$(B_{YE} - 1) E_{S,YE}^0 ,$$
(1.50)

in analogy with (1.45) using the definition

$$B_{YE} = \frac{E_{YE} + E_{S,YE}^0 - E_{YE}^0}{E_{S,YE}^0} .$$
(1.51)

1.4.3 Proximity Energy

The proximity model [Błocki 77], [Błocki 81] accounts for the nuclear interaction energy between two nuclei by relating it to the surface energy. It is equivalent to the proximity theorem which states that *the force is proportional to the interaction potential per unit area*, $e(D)$, *between two flat surfaces*. Surface integration over the area of the gap or crevice of width D yields the proximity energy (which is negative by convention)

$$E_{\text{prox}} = \int dS \; e(D).$$
(1.52)

For axially symmetric systems where each fragment has the same surface width a ($\approx$ 1fm) this simplifies [Feldmeier 80] to a one dimensional integral over the transverse radial distance ϱ, cf. Fig. 1.1,

$$E_{\text{prox}} = 4\pi\sigma \int\limits_{\varrho_{\min}}^{\varrho_{\max}} d\varrho\varrho \; \varphi \left(\frac{D(\varrho)}{a} \right),$$
(1.53)

where D is the the longitudinal gap dimension and $\varrho_{\min}, \varrho_{\max}$ are the radial

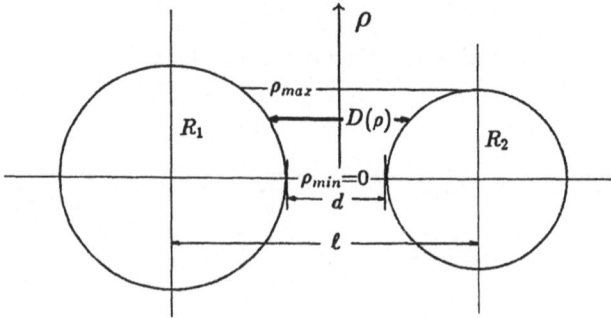

Figure 1.1: Coordinates and parameters in the proximity model.

locations where the gap is minimal or maximal, respectively, ($\varrho_{\min} = 0$ for separated shapes and $\varrho_{\max}$ may be extended to infinity if $\varphi(x)$ vanishes sufficiently fast); $\varphi(x) = e(ax)/2\sigma$ is a universal function (see Fig. 9.2) resulting from the choice of the interaction potential, and σ is the surface tension. In [Błocki 77] the Seyler-Blanchard force was employed [Seyler 61] yielding the universal function which is discussed in Sect. 9.1. In units of E_S^0 we define

$$B_{\text{prox}} = \frac{E_{\text{prox}}}{E_S^0} = \frac{1}{R_0^2} \int_{\varrho_{\min}}^{\varrho_{\max}} d\varrho \, \varrho \, \varphi\left(\frac{D(\varrho)}{a}\right) . \tag{1.54}$$

The proximity energy is used mainly as the real part of the heavy ion interaction potential, but it can also be thought of as a correction to the surface energy similar to the Krappe-Nix or Folded-Yukawa energy, see Sect. 1.4.2.

1.5 Coulomb Energies

The calculation of the Coulomb energy (positive) or gravitational energy (negative) of extended objects has engaged innumerable scientists and mathematicians over the years. The general expressions are given below and most of the specific results are referred to later in the appropriate sections.

1.5.1 Direct

The most general expression for the Coulomb energy of a nucleus is

$$E_{\text{Coul}} = \frac{1}{2} \int d^3 r \, \varrho(\boldsymbol{r}) V(\boldsymbol{r}) , \tag{1.55}$$

where $\varrho(\boldsymbol{r})$ is the proton particle number density and $V(\boldsymbol{r})$ is the potential (in MeV) felt by a particle of charge e, given by the expression

$$V(\boldsymbol{r}) = e^2 \int d^3 r' \varrho(\boldsymbol{r}') \mid \boldsymbol{r} - \boldsymbol{r}' \mid^{-1} . \tag{1.56}$$

For a uniform spherical charge distribution of radius R_0

$$V(r) = Ze^2 \begin{cases} \dfrac{1}{R_0} \left[\dfrac{3}{2} - \dfrac{1}{2} \left(\dfrac{r}{R_0} \right)^2 \right] & , r \leq R_0 \\[4mm] \dfrac{1}{r} & , r \geq R_0 \end{cases} \tag{1.57}$$

Here, and elsewhere in this book, the quantity e^2 appears representing the square of the charge of an electron (or proton). Its value [Aguilar 86], [Robinson 87] is

$$e^2 = 1.43996518 \, (44) \text{ MeV fm} . \tag{1.58}$$

For a uniform spherical charge distribution of radius R_0

$$E_C^0 = \frac{3}{5} \frac{e^2 Z^2}{R_0} , \tag{1.59}$$

and for an arbitrary distribution the energy can be expressed in terms of this quantity by the expression

$$E_{\text{Coul}} = \frac{15}{32\pi^2} E_C^0 \int d\tau \int d\tau' \ |\boldsymbol{r} - \boldsymbol{r}'|^{-1} , \tag{1.60}$$

where $d\tau$ is the volume element in units of R_0^3 and the vectors $\boldsymbol{r}$ and $\boldsymbol{r}'$ are in units of R_0. It is often convenient to work with the relative Coulomb energy defined by the expression

$$B_{\text{Coul}} = E_{\text{Coul}}/E_C^0 . \tag{1.61}$$

1.5.2 Exchange

The exchange contribution to the Coulomb energy is a quantum phenomenon and can be correctly calculated only from the corresponding proton wave functions. However, the commonly employed Fermi gas approximation [Slater 60] results in the expression

$$E_{\text{ex}} = -e^2 \frac{3}{4} \left(\frac{3}{\pi}\right)^{1/3} \int dV \ \varrho(\boldsymbol{r})^{4/3} , \tag{1.62}$$

where $\varrho(\boldsymbol{r})$ is the local proton particle number density. For a uniform spherical charge distribution of radius R_0 this expression becomes

$$E_{\text{ex}}^0 = -\frac{3}{4} \left(\frac{3}{2\pi}\right)^{2/3} \frac{e^2 Z^{4/3}}{R_0} . \tag{1.63}$$

For volume preserving deformations of uniform charge distributions there is no shape dependence for (1.62) analogous to (1.60) and the exchange correction is essentially a volume term.

1.6 Curvature and Redistribution Energies, etc

In extensions of the LDM one term frequently encountered is the *curvature energy*. This quantity is proportional to the integrated curvature and is thought of as a correction to the surface energy term. The local curvature κ is defined by the expression

$$\kappa = \frac{1}{2} \left(R_{\text{min}}^{-1} + R_{\text{max}}^{-1} \right) , \tag{1.64}$$

where R_{min} and R_{max} are the two principal radii of curvature at the local point, and the integrated curvature is simply

$$K = \int dS \ \kappa . \tag{1.65}$$

For a sphere of radius R_0 the integrated curvature is

$$K_{\text{sphere}} = 4\pi R_0 . \tag{1.66}$$

Under the usual LDM assumption of incompressibility (1.10) such a term can be written as

$$E_{\text{curv}} = c_{\text{curv}} A^{1/3} B_{\text{curv}} , \qquad (1.67)$$

where the coefficient c_{curv} is determined from measured quantities (such as nuclear masses) or estimated on the basis of theoretical models. The $A^{1/3}$ dependence arises from the fact that the local curvature of a sphere of radius R_0 is simply $1/R_0$ and the integrated curvature is $4\pi R_0$. The shape dependence of (1.67) is, as usual, contained in the quantity B_{curv}, the relative curvature energy, which can be expressed in terms of the curvature energy of the sphere, $E^0_{\text{curv}} = c_{\text{curv}} A^{1/3}$, by the dimensionless surface integral

$$B_{\text{curv}} = \frac{E_{\text{curv}}}{E^0_k} = \frac{1}{8\pi} \int ds \left(r^{-1}_{\min} + r^{-1}_{\max} \right) , \qquad (1.68)$$

where $r_{\min}, r_{\max}$ are the curvature radii of the surface element in units of R_0 .

The Gaussian curvature Γ (mentioned in [Błocki 77] in connection with higher order terms in the leptodermous expansion) is defined by the expression

$$\Gamma = R^{-1}_{\min} R^{-1}_{\max} . \qquad (1.69)$$

The integrated Gaussian curvature for any closed surface with the topology of a sphere is equal to 4π and a term in a LDM energy expression proportional to this quantity would be a constant independent of mass number or shape (except when the topology undergoes a change, as in fission).

The relative compression energy, which is referred to in [Hasse 71], is just the square of the relative surface energy,

$$B_{\text{comp}} = B^2_{\text{surf}} . \qquad (1.70)$$

The redistribution energy and surface redistribution energies of first and second kind in the Droplet Model, see Sect. 1.1.2 and [Myers 69], are mean squared surface or volume deviations of the Coulomb potential Φ in units of Ze^2/R_0, where Φ is equal to V of Eq. (1.56),

$$\begin{aligned}
B_{\text{red}} &= \frac{175}{4\pi} \int d\tau \left(\Phi - \overline{\Phi} \right)^2 \\
B_{\text{sr1}} &= \frac{25}{16\pi^2} \left[\int ds \left(\Phi_s - \overline{\Phi} \right) \right]^2 \\
B_{\text{sr2}} &= \frac{25}{4\pi} \int ds \left(\Phi_s - \overline{\Phi} \right)^2 .
\end{aligned} \qquad (1.71)$$

Other notations [Myers 69]:

$$\begin{aligned}
B_{\text{v}} &= \sqrt{B_{\text{sr1}}} \\
B_{\text{w}} &= B_{\text{sr2}} .
\end{aligned} \qquad (1.72)$$

Here Φ is defined in the volume, Φ_s is its surface value, $\overline{\Phi}$ and $\overline{\Phi}_s$ are the volume and surface averages, respectively, and the following interrelations hold

$$B_{\text{red}} = \frac{175}{3}\left(\overline{\Phi^2} - \overline{\Phi}^2\right)$$

$$B_{\text{sr1}} = (5\overline{\Phi}_s)^2 - 60\overline{\Phi}_s B_{\text{Coul}} B_{\text{surf}} + (6B_{\text{Coul}} B_{\text{surf}})^2$$

$$B_{\text{sr2}} = 25\overline{\Phi_s^2} - 60\overline{\Phi}_s B_{\text{Coul}} + 36B_{\text{Coul}}^2 B_{\text{surf}}$$

$$\overline{\Phi} = \frac{6}{5}B_{\text{Coul}}. \tag{1.73}$$

1.7 Deformation Energies

In the pure LDM with surface and Coulomb energies only (the volume energy is independent of deformation) the total deformation energy is defined as the actual total energy minus the value for a sphere,

$$E_{\text{Def}} = E_{\text{surf}} - E_S^0 + E_{\text{Coul}} - E_C^0 . \tag{1.74}$$

In units of the surface energy of the sphere and with help of the fissility [Bohr 39]

$$x = \frac{E_C^0}{2E_S^0} = \frac{Z^2/A}{2a_S/c_1}$$

$$= \frac{Z^2/A}{(Z^2/A)_{\text{crit}}} \approx \frac{Z^2/A}{50} \tag{1.75}$$

it becomes

$$B_{\text{Def}} = (B_{\text{surf}} - 1) + 2x(B_{\text{Coul}} - 1). \tag{1.76}$$

The fissility x is defined in such a way that $x \ll 1$ corresponds to light nuclei with a fission barrier shape close to two tangent spheres and $x \approx 1$ to heavy nuclei with a fission barrier shape close to sphericity. Other energies like the curvature, compressibility or redistribution energies can be treated on the same footing, cf. [Hasse 71], by defining parameters E_k^0/E_S^0, etc. analogous to the fissility.

In studying the stability against rotations, the rotational energy

$$E_{\text{rot}} = \frac{L^2}{2\mathcal{J}} \tag{1.77}$$

is added where L is the angular momentum and $\mathcal{J}$ is the moment of inertia around a suitable axis.

Introducing the parameter

$$z = \frac{E_{\text{rot}}}{E_S^0} = \frac{L^2}{2\mathcal{J}_0 E_S^0} \tag{1.78}$$

the relative rotational energy simply becomes

$$B_{\text{rot}} = \frac{\mathcal{J}_0}{\mathcal{J}} \tag{1.79}$$

and the total relative deformation energy reads

$$B_{\text{Def}} = (B_{\text{surf}} - 1) + 2x(B_{\text{Coul}} - 1) + z(B_{\text{rot}} - 1) \,. \tag{1.80}$$

z is usually denoted by y which, however, can cause some confusion with the quantity $y = 1 - x$ discussed in Sect. 7.5.

1.8 Normal Modes and Dynamics

For a given set of dimensionless shape coordinates $\{\alpha\}$ the deformation energy can be expanded around a stationary point $\{\alpha^0\}$, for instance the ground state or the saddle point,

$$B_{\text{Def}}(\alpha) = \frac{1}{2} \sum_{ij} C_{ij} (\alpha_i - \alpha_i^0)(\alpha_j - \alpha_j^0), \tag{1.81}$$

which defines the dimensionless stiffness parameters

$$C_{ij} = \frac{\partial^2 B_{\text{Def}}(\alpha)}{\partial \alpha_i \partial \alpha_j} \bigg|_{\alpha = \alpha^0} \,. \tag{1.82}$$

The kinetic energy

$$E_{\text{kin}}(\alpha, \dot\alpha) = \frac{1}{2} M R_0^2 \sum_{ij} B_{ij}(\alpha) \dot\alpha_i \dot\alpha_j \tag{1.83}$$

defines the mass (or inertia) parameters B_{ij} in units of the total mass M times R_0^2. Similarly, the rotational energy for rotations about the axes $k = 1, 2, 3$ with angular velocities ω_k is written as

$$E_{\text{rot}}(\omega, \alpha) = \frac{1}{2} M R_0^2 \sum_{k} B_{\omega_k}(\alpha) \dot\omega_k^2 \,. \tag{1.84}$$

They are to be calculated in some model, for instance the hydrodynamic model [Wilets 64], [Kelson 64], [Nix 67], [Hasse 68a], [Hasse 75]. If stiffness and mass parameters are diagonal the eigenfrequencies Ω_i are also diagonal,

$$\omega_i = \Omega_i \Big/ \sqrt{\frac{E_s^0}{M R_0^2}} = \sqrt{\frac{C_{ii}}{B_{ii}}} \,, \tag{1.85}$$

where ω_i are the corresponding dimensionless quantities, otherwise they have to be calculated by diagonalization.

Dissipation can be incorporated into the dynamics of a liquid drop via ordinary two body viscosity. If η denotes the viscosity constant, the (two body) viscosity coefficients Z_{ij} are defined through the time rate of change of the energy loss,

$$- \dot W_{\text{visc}}(\alpha, \dot\alpha) = 4\pi\eta R_0^3 \sum_{ij} Z_{ij}(\alpha) \dot\alpha_i \dot\alpha_j. \tag{1.86}$$

The one-body dissipation coefficients D_{ij}, on the other hand, [Błocki 78] are defined similarly through

$$- \dot W_{\text{one body}}(\alpha, \dot\alpha) = 4\pi\varrho_m \bar v R_0^4 \sum_{ij} D_{ij}(\alpha) \dot\alpha_i \dot\alpha_j, \tag{1.87}$$

15

where ϱ_m is the mass density, and $\bar{v} = 3v_F/4$ is the average particle velocity (v_F is the Fermi velocity) in the Fermi gas model. For a discussion of a smaller value of $\bar{v}$ cf. [Griffin 86].

Chapter 2

Characterization of Leptodermous Distributions

Almost all of the early work in macroscopic (Liquid Drop Model) nuclear physics idealized the nucleus as an object (spherical or deformed) with a constant uniform density inside a sharp surface. This is possible, of course, because the nuclear density is fairly uniform in the interior and the falloff of the density in the surface is localized (for heavy nuclei at least) to a region that is thin compared to the size of the system.

Many different algebraic functions have been employed to represent nuclear density distributions and potential wells. The ubiquitous Fermi function of Sect. 4.2.1.1 is the main example, but many other functions have been used. Collections of alternative expressions can be found in [Budzanowski 76] and [Collard 67]. Since a number of different functional forms may be equally acceptable for the interpretation of a particular experiment various approaches have been developed for the characterization of these expressions in terms of *abstract* quantities which are independent of the particular functions used. The discussion that follows is based mainly on [Myers 73] and the formal extension of that work in [Süssmann 75].

2.1 Introduction

For spherically symmetric distributions with a well established uniform central region and a falloff to zero which is localized in the surface, the radial geometrical properties can be described in terms of the following quantities

$$C, \quad \text{the } \textit{central radius,}$$
$$R, \quad \text{the } \textit{equivalent sharp radius}$$
$$Q, \quad \text{the } \textit{equivalent rms radius} \text{ and}$$
$$b, \quad \text{the } \textit{surface width} . \tag{2.1}$$

The central radius C and the surface width b are the integral counter-parts of the punctual quantities:

$$R_{\frac{1}{2}}, \quad \text{the } \textit{half value radius,}$$
$$t_{10-90}, \quad \text{the } \textit{10} - \textit{90\% distance} . \tag{2.2}$$

They are defined in terms of linear moments (as opposed to spherical moments) of the normalized *distribution function* $f(r)$. The functions we are concerned with here are all spherically symmetric and fall monotonically and continuously from

$$f(0) = 1 \quad \text{to} \quad f(\infty) = 0 \,. \tag{2.3}$$

These boundary conditions permit us to define a *weight function*

$$g(r) := -\frac{\mathrm{d}f(r)}{\mathrm{d}r} \,, \tag{2.4}$$

which is automatically normalized to unity

$$\int_0^\infty g(r)\,\mathrm{d}r = 1 \,. \tag{2.5}$$

The quantities C and b are then given by the first two moments of $g(r)$,

$$C := \int_0^\infty g(r)\,r\,\mathrm{d}r \,, \tag{2.6}$$

$$b^2 := \int_0^\infty g(r)(r - C)^2\,\mathrm{d}r \,. \tag{2.7}$$

In a similar way additional information about the surface is available from higher moments such as the skewness and kurtosis of the distribution which can be obtained from the quantities Γ_3 and Γ_4 defined below in Sect. 2.3.

An alternative expression for (2.6) can be given in terms of the norm of $f(r)$

$$F := \int_0^\infty f(r)\,r^2\,\mathrm{d}r \,, \tag{2.8}$$

and the *charge moments* F_n defined by

$$F_n := \int_0^\infty f(r)\,r^n\,\mathrm{d}r \ = F\,\langle r^{n-2}\rangle \,, \tag{2.9}$$

where the bracket $\langle\ \rangle$ denotes the *charge mean* $\langle \Psi(r)\rangle$ of any continuous function $\Psi(r)$ which is defined by the expression

$$\langle \Psi(r)\rangle := F^{-1} \int_0^\infty \Psi(r)\,f(r)\,r^2\,\mathrm{d}r \,. \tag{2.10}$$

The quantity C in Eq. (2.6) above can then be expressed as

$$C := \int_0^\infty f(r)\,\mathrm{d}r \ = F_0 \ = F\,\langle r^{-2}\rangle \,, \tag{2.11}$$

for which it is also true that

$$\int_0^\infty [f(r) - \theta(C - r)]\,\mathrm{d}r = 0 \,. \tag{2.12}$$

18

The next quantity of interest is the *equivalent sharp radius R*. It can be defined as the radius of a uniform sharp distribution having the same value in the bulk and the same volume integral as $f(r)$, i.e.,

$$\frac{4}{3}\pi R^3 f(\text{bulk}) := 4\pi \int_0^\infty f(r)\, r^2 \, dr \; . \tag{2.13}$$

For smooth leptodermous distributions (such as a Fermi distribution) the bulk value corresponds very closely to the central value and to this approximation $f(0)$ can be substituted for $f(\text{bulk})$ in (2.13). Of course, for distributions that are leptodermous except for some smooth oscillations in the interior (for example, nuclear density distributions found in shell model or Hartree-Fock calculations) the punctual value $f(0)$ is clearly inadequate for the definition of R and some suitable average bulk value must be employed.

If we seek an approach similar to (2.12) for the definition of the *equivalent sharp radius R* we can write

$$\int_0^\infty [f(r) - \theta(R - r)]\; r^2 \, dr = 0 \; , \tag{2.14}$$

or from (2.8) or (2.9)

$$\frac{1}{3}\, R^3 := \int_0^\infty f(r)\, r^2 \, dr \; = F_2 \; = F \; . \tag{2.15}$$

From (2.14) we see that R is the radius of a sharp surface sphere having the same volume integral as $f(r)$. In a similar way the *equivalent rms radius Q* can be defined by

$$\frac{3}{5}\, Q^2 := \langle r^2 \rangle = \frac{F_4}{F_2} \; , \tag{2.16}$$

where generally

$$\langle r^k \rangle = \frac{F_{k+2}}{F_2} \; . \tag{2.17}$$

The quantity Q is seen to be the radius of a sharp surface sphere that has the same rms radius as $f(r)$.

The quantity Q is the special case for $k = 2$ of the quantity

$$R_k := \left[\frac{1}{3}(k + 3)\langle r^k \rangle\right]^{1/k} \; , \tag{2.18}$$

of Ford and Wills [Ford 69], and it is related to the quantity

$$M(k) := \langle r^k \rangle^{1/k} \tag{2.19}$$

employed by Friedrich and Lenz [Friedrich 72] for the characterization of charge distributions obtained from electron scattering measurements. We might also mention in passing that the *generalized radial moments* defined by Ford and Rinker [Ford 73]

$$\langle u(r)\, r^k \rangle := \frac{4\pi}{Ze} \int_0^\infty \varrho(r)\, u(r)\, r^{k+2} \, dr \; , \tag{2.20}$$

19

where

$$Ze = 4\pi \int_0^\infty \varrho(r)\, r^2 \mathrm{d}r \,, \tag{2.21}$$

are sometimes employed in studies of muonic atoms. In particular the expression

$$u(r) = c + dr^k e^{\alpha r} \tag{2.22}$$

has been advocated in [Barret 70] on physical grounds.

Of the three quantities C, R and Q, the quantity of fundamental geometric importance is the equivalent sharp radius R. A sharp sphere having this radius is in a basic sense representative of the distribution $f(r)$. If the uniform central density of such a sharp sphere is set equal to the bulk value of $f(r)$, as defined in connection with (2.13), then this sphere has the same volume integral as $f(r)$ and it differs from $f(r)$ *only in the surface region* (namely in the degree of diffuseness). The quantity C is mainly of interest because $R_{\frac{1}{2}} = C$ for the symmetric surface functions (such as Fermi distributions) often employed to characterize nuclear densities and potential wells. In addition, C provides a more precise description than does R for the *location* of the surface that enters into practical applications of the proximity method of Sect. 1.4.3. The *equivalent rms radius* Q is of interest since it is expected that this is the property of the distribution that is measured in some experiments. As can be seen in Fig. 2.1, sharp spheres with the same volume integral as $f(r)$ having the radii C or Q grossly misrepresent the appearance of the function $f(r)$, since they substantially differ from it over the bulk region.

Süssmann has given exact expressions relating C and Q to R in terms of b and higher order moments of the surface and these are given in the following sections that are based on [Süssmann 75]. However, the following approximate expressions suffice for many applications, and serve as simple reminders of the relationships of these quantities to each other:

$$C = R(1 - \beta^2 \cdots) \,, \tag{2.23}$$

$$Q = R\left(1 + \frac{5}{2}\beta^2 \cdots\right) \,, \tag{2.24}$$

where

$$\beta := b/R \,. \tag{2.25}$$

These relationships hold for $\beta \ll 1$ in the general spirit of this chapter where only leptodermous systems are being considered. An interesting extension of these ideas is contained in [Viñas 75] where the quantities C, R and β are redefined in a way that lends itself to the description of distributions where β is not small or where oscillations in the interior of the distribution make it difficult to apply the expressions presented here.

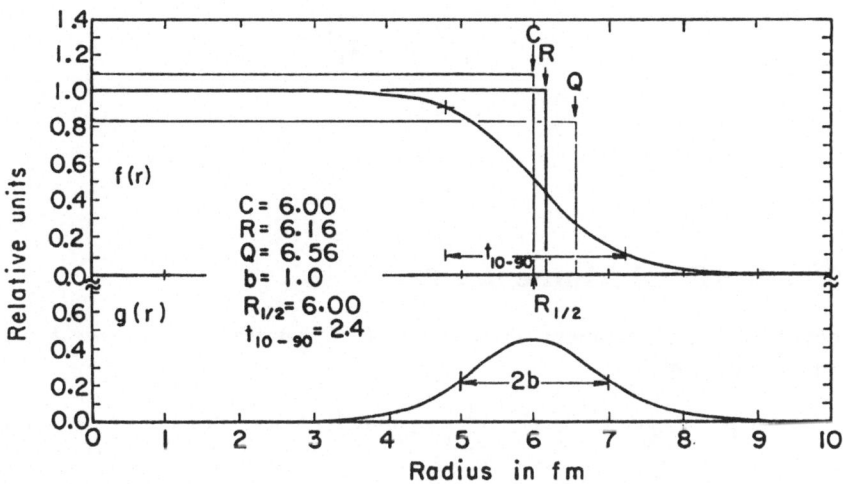

Figure 2.1: The normalized, spherically symmetric, leptodermous distribution $f(r)$ and the corresponding surface distribution function $g(r)$ for a hypothetical nucleus are plotted against the radial distance r. The values of $R_{1/2}$ and t_{10-90} are given for this distribution in addition to the values of C, R, Q, and b. Sharp density distributions having the same volume integral as $f(r)$ and radii equal to C, R, and Q have also been drawn in for the purpose of demonstrating the geometrical importance of R.

2.2 The Original Surface Moments, G_m

In analogy with (2.11) we can define a *surface mean* of the function $\Psi(r)$ by the expression

$$\{\Psi(r)\} := \int_0^\infty \Psi(r)\, g(r)\, \mathrm{d}r \, , \tag{2.26}$$

where the automatically normalized *weight function* $g(r)$ is defined in (2.4) above. Partial integration yields the identity,

$$\{\Psi(r)\} = \int_0^\infty f(r)\, \Psi'(r)\, \mathrm{d}r = F \langle r^{-2}\, \Psi'(r) \rangle \, . \tag{2.27}$$

The *original surface moments* G_m are defined by

$$G_m := \int_0^\infty g(r)\, r^m\, \mathrm{d}r = \{r^m\} \, . \tag{2.28}$$

Of course, the normalization integral

$$G = G_0 = 1 \, , \tag{2.29}$$

and

$$G_m = m\, F_{m-1} \, . \tag{2.30}$$

Other relationships of possible interest are

21

$$\langle r^k \rangle = \frac{3}{k+3} \frac{\{r^{k+3}\}}{\{r^3\}} \tag{2.31}$$

$$C = G_1 \tag{2.32}$$

$$R^3 = G_3 \tag{2.33}$$

$$Q^2 = G_5/G_3 \ . \tag{2.34}$$

2.3 The Surficial Moments, Γ_n

Moments of the surface *weight function* $g(r)$ (2.4) relative to the *central radius* C can be defined by

$$\Gamma_n := \{(r - C)^n\} = \sum_{m=0}^{n} (-1)^{n-m} \binom{n}{m} C^{n-m} G_m \ , \tag{2.35}$$

which can be inverted to yield

$$G_m = \sum_{n=0}^{m} \binom{m}{n} C^{m-n} \Gamma_n \ . \tag{2.36}$$

These *surficial moments* Γ_n have the properties that

$$\Gamma_0 = G_0 = 1 \ , \tag{2.37}$$

$$\Gamma_1 = 0 \tag{2.38}$$

and the *surface width* b of (2.7) is, of course, given by

$$b^2 := \Gamma_2 = G_2 - G_1^2 \tag{2.39}$$

Other relations of interest are:

$$R^3 = C^3 + 0 + 3C\Gamma_2 + \Gamma_3 \tag{2.40}$$

$$R^3 Q^2 = C^5 + 0 + 10C^3 \Gamma_2 + 10C^2 \Gamma_3 + 5C\Gamma_4 + \Gamma_5 \ . \tag{2.41}$$

2.4 The Surface Shape Coefficients, γ_n

The quantity C locates the surface and b is a measure of its size (or extent). The *shape* of the surface influences the higher moments, which can be written in dimensionless form as

$$\gamma_n := b^{-n} \Gamma_n \ . \tag{2.42}$$

If, in the leptodermous case, we replace Γ by γ in the expressions (2.40) and (2.41) and expand in powers of the *skin coefficient*

$$\beta := b/R \ll 1 \ , \tag{2.43}$$

we obtain the relationships

22

$$C = R\left(1 - \beta^2 - \frac{1}{3}\gamma_3\beta^3 + 0 - \frac{1}{3}\gamma_3\beta^5 \cdots\right) \tag{2.44}$$

$$Q = R\left[1 + \frac{5}{2}\beta^2 + \frac{25}{6}\gamma_3\beta^3 + \frac{5}{2}\left(\gamma_4 - \frac{21}{4}\right)\beta^4 + \frac{1}{2}\left(\gamma_5 - \frac{275}{6}\gamma_3\right)\beta^5 \cdots\right] \tag{2.45}$$

which were referred to earlier in connection with Eqs. (2.23, 2.24). In [Süssmann 75] it is pointed out that Eq. (2.45) is essentially the same as one given in [Elton 61] for the special case of a Fermi function.

2.5 Distributions Related by Folding

There is another class of geometrical relationships that is also of considerable interest. These relationships connect the geometrical properties of one lepto-dermous distribution to the corresponding properties of a second distribution which is obtained from the first by folding in a function of short range. One example is a nuclear charge distribution obtained by folding the proton charge distribution into the assumed spatial distribution of the protons. Another example is a single-particle or optical-model potential well obtained by folding a two-body interaction into the nuclear density distribution. The general properties of such distributions are discussed in Chap. 3, but some of the expressions that apply specifically to spherically symmetric leptodermous distributions can be found here.

Probably the best-known expression for the characterization of a distribution obtained by folding is (3.4). In the notation of (2.17) this becomes

$$\langle r^2 \rangle = \langle r^2 \rangle_g + \langle r^2 \rangle_f \tag{2.46}$$

for a generating function $g(r)$ and a folding function $f(r)$. Of course, (2.46) can be combined with the definition of Q in (2.16) to give

$$Q^2 = Q_g^2 + \frac{5}{3}\langle r^2 \rangle_f . \tag{2.47}$$

Another useful relationship is

$$b^2 = b_g^2 + b_f^2 + \text{terms of order } \beta^2 , \tag{2.48}$$

where the *width* of the folding function $f(r)$ has the special definition

$$b_f^2 := \frac{1}{3}\langle r^2 \rangle_f \tag{2.49}$$

in terms of the radial moment $\langle r^2 \rangle$ defined by (2.17). Eq. (2.48) relates the surface width b, of a generating function g to the width b of the distribution that results from folding. Of course, for a uniform generating function with a sharp surface b_g is zero and then (to second order in the expansion parameter β from (2.25)) the width of the distribution is just that of the folding function b_f. For the three fundamental quantities C, R and Q the following relationships hold for $\beta_f \ll 1$:

$$C = C_g(1 - \beta_f^2 \cdots)$$
$$R = R_g$$
$$Q = Q_g\left(1 + \frac{5}{2}\beta_f^2 \cdots\right) , \tag{2.50}$$

where

$$\beta_f := b_f/R \,. \tag{2.51}$$

These expressions show that when a short-range function is folded into a lepto-dermous distribution, another distribution is obtained that has a larger equivalent rms radius Q, an identical equivalent sharp radius R, and a central radius C that is smaller than the values of the corresponding quantities for the initial distribution.

Chapter 3

Folded Distributions

The spatial distribution of the nuclear density (or the nuclear single-particle potential) can be roughly approximated by a uniform sphere with a sharp surface and a volume proportional to the number of nucleons. When this simplest approach is generalized to include other shapes and a diffuse surface the mathematical difficulties associated with calculating various properties of these distributions can become formidable. Indeed, the subject forms the basis of this book. In Chap. 4 many examples of spherically symmetric distributions with diffuse surfaces are considered. In Chaps. 5 through 10 a wide variety of other shapes are considered but only for uniform distributions with a sharp surface. It is not widely appreciated that a uniform distribution can conveniently be generalized to one with a diffuse surface by folding into it a suitable short ranged function.

The idea of folding a short ranged function into a uniform distribution to create a diffuse surface for describing nuclear density distributions seems to have originated with [Helm 56] and this early work has been brought up to date by [Friedrich 82]. The geometrical features of such folded distributions which are discussed in more detail in the remainder of this chapter, were introduced by [Myers 73] and considered in [Myers 76] and [Krappe 76]. However, the definitive work on the subject is [Krappe 81] and this reference should be consulted by any reader seriously interested in this approach. A general discussion of the calculation of moments, potentials, and energies for arbitrarily shaped distributions with diffuse surfaces created by folding is also contained in [Davies 76a].

The idea of creating a single particle or optical model potential well by folding a simple phenomenological direct two body force into an assumed density distribution enjoyed a period of popularity. That this approach is likely to be unsatisfactory can be seen in [Myers 70] and the geometric reasons underlying the disagreements that arise are discussed in [Myers 76]. More recently the geometrical properties have been considered of potential wells generated by using folding potentials whose strength is density dependent, two examples can be found in [Srivastava 82a] and [Srivastava 83].

3.1 Definition

As it is used here the folding product $g * f$ of two functions $g(\boldsymbol{r})$ and $f(s)$ is defined by

$$g(\boldsymbol{r}) * f(s) = \int \mathrm{d}^3 r' \, g(\boldsymbol{r}) \, f(s) \,, \tag{3.1}$$

where
$$s = |\boldsymbol{r} - \boldsymbol{r}'| \, .$$

It is generally convenient to consider only normalized folding functions

$$\int d^3r \, f(r) = 1 \tag{3.2}$$

and uniform generating functions $g(\boldsymbol{r})$ which are completely characterized by the location of the surface.

3.2 Normalization and Radial Moments

The radial moments m_n of a distribution $f(r)$ are defined by Eq. (1.14) and for a convolution (3.1) of two functions g and f the normalization is

$$m_0 = m_0^{(g)} m_0^{(f)} \, , \tag{3.3}$$

which means that for normalized folding functions the volume integral of the diffuse distribution is the same as the volume integral of the generating distribution. No special scaling or other shape dependent correction factors are required to maintain the normalization as in the case of distributions which are made diffuse by other means (a Fermi function (4.15) for example). The second moments are additive so that

$$m_2 = m_2^{(g)} + m_2^{(f)} \, , \tag{3.4}$$

which is just (2.46) in another notation.

3.3 Multipole Moments

In [Myers 76] and [Krappe 81] it is shown that the multipole moments (see Sect. 1.3.3) of a diffuse distribution formed from the convolution of a generating function g and a normalized spherically symmetric folding function f are identical with the moments of g itself, i.e.

$$M_{\ell m} = M_{\ell m}^{(g)} \, , \tag{3.5}$$

and in particular

$$Q_\ell = Q_\ell^{(g)} \, . \tag{3.6}$$

This identity means that only sharp surface distributions need be employed in the discussion of physical phenomena that are concerned with multipole moments.

3.4 Moments of Inertia

If convolution is used to create the diffuseness, then the moments of inertia (see Sect. 1.3.5) of the diffuse and sharp distributions are related by the simple expression

$$I = I_g + 2Mb_f^2 , \tag{3.7}$$

where I is the moment of inertia of the diffuse distribution about an arbitrary axis, I_g is the corresponding moment of inertia for the generating shape. The quantity M is the total mass of the object and b_f is the *width* of the folding function defined by (2.49).

3.5 Coulomb Energy

General expressions for the calculation of Coulomb energies of diffuse distributions created by folding can be found in [Krappe 76] and [Davies 76a]. The analytic result for a spherical distribution and a Yukawa folding function is given in (4.68).

3.6 Specific Examples

The most commonly employed folding functions are the spherical square well, the Yukawa, the Gaussian and the Yukawa-plus-exponential.

For a **square well**

$$f_S(r) = \left(\frac{4}{3}\pi a^3\right)^{-1} \Theta(r - a) \tag{3.8}$$

the radial moments are

$$m_n = a^n \frac{3}{(n+3)} \tag{3.9}$$

and the *width* of (2.49) is

$$b_S = a\sqrt{\frac{3}{10}} . \tag{3.10}$$

For a **Yukawa** function

$$f_Y(r) = (4\pi a^3)^{-1} \frac{e^{-r/a}}{r/a} \tag{3.11}$$

the radial the moments are

$$m_n = a^n(n+1)! \tag{3.12}$$

and the *width* (2.49) is

$$b_Y = a\sqrt{2} . \tag{3.13}$$

For a **Gaussian** distribution

$$f_G(r) = (a\sqrt{\pi})^{-3} e^{-r^2/a^2} \tag{3.14}$$

the radial moments are

$$m_n = a^n \begin{cases} \dfrac{(n+1)!}{(n/2)!\,2^n} & , \; n \text{ even} \\[2ex] \dfrac{2}{\sqrt{\pi}} \left[\dfrac{1}{2}(n+1) \right]! & , \; n \text{ odd} \end{cases} \tag{3.15}$$

and the *width* (2.49) is

$$b_G = a/\sqrt{2} . \tag{3.16}$$

For Yukawa and Gaussian folding functions the convolution with a uniform spherical generating function

$$g(r) = g_0 \, \Theta(R - r) \tag{3.17}$$

can be done analytically. The corresponding expressions are discussed in Sect. 4.2.4.

The **Yukawa-plus-exponential** distribution

$$f_{\text{YE}}(r) = (r/a - 2) \frac{e^{-r/a}}{r/a} \tag{3.18}$$

is discussed in Sect. 1.4.2 in connection with the generalization of the surface energy integral to a double folding of an interaction potential over the assumed nuclear density distribution. This distribution (3.18) is especially designed to have positive and negative parts that cancel exactly, consequently the normalization integral $m_0 = 0$ and all the other moments

$$m_n = a^n n(n+1)! \tag{3.19}$$

are only defined to within a constant factor.

Chapter 4

Spherically Symmetric Distributions

Many different functions have been considered for representing the distribution of nuclear mass and charge, or the distribution of the nuclear single particle and optical model potential wells.

4.1 Sharp Sphere of Radius R

4.1.1 Uniform Density

For a uniform spherically symmetric distribution of charge or mass of radius R_0 some of the elementary geometrical properties are (for the curvatures see Sect. 1.6),

$$
\begin{aligned}
\text{Volume} &= 4\pi R_0^3/3 \\
\text{Surface} &= 4\pi R_0^2 \\
\text{Integrated mean curvature} &= 4\pi R_0 \\
\text{Integrated Gaussian curvature} &= 4\pi \, .
\end{aligned}
\tag{4.1}
$$

In addition, the moment of inertia about an axis through the center is (1.28)

$$
J_0 = \frac{2}{5} M R_0^2 \, ,
\tag{4.2}
$$

where M is the total mass. The Coulomb potential for a uniformly charged sphere is given in (1.57) and the Coulomb energy is (1.59)

$$
E_{\text{Coul}}^0 = \frac{3}{5} \frac{Q^2}{R_0} \, ,
\tag{4.3}
$$

where $Q = eZ$ is the total charge.

4.1.2 Central Depression

Two different ways to describe a spherically symmetric distribution with a sharp surface located at R_0 and a quadratic central depression that have been used are:

$$
\text{Type I} : \quad \varrho = \varrho_0 \left[1 + \alpha \, f(r) \right] ,
\tag{4.4}
$$

where

$$
f(r) = -\frac{3}{5} + \left(\frac{r}{R_0} \right)^2 ,
$$

or

$$\text{Type II}: \quad \varrho = \varrho_0 \left(1 + \frac{3}{5}\omega\right)^{-1} \left[1 + \omega \left(\frac{r}{R_0}\right)^2\right] . \tag{4.5}$$

The relationship between α and ω is

$$\left(1 - \frac{3}{5}\alpha\right)\left(1 + \frac{3}{5}\omega\right) = 1 . \tag{4.6}$$

If either α or ω is small then $\alpha \approx \omega$.
The effective volume, and the surface properties are still given by (4.1).

Coulomb Energy for expressions of Type II

    Direct (see Sect. 1.5.1 and [Lindner 69], [Myers 69])

$$E_{\text{Coul}} = E^0_{\text{Coul}} \frac{1 + \frac{8}{7}\omega + \frac{1}{3}\omega^2}{(1 + \frac{3}{5}\omega)^2} . \tag{4.7}$$

    Exchange (see Sect. 1.5.2 and [Lindner 69])

$$E_{\text{ex}} = E^0_{\text{ex}} \frac{1 + \frac{4}{5}\omega + \frac{2}{21}\omega^2}{(1 + \frac{3}{5}\omega)^{4/3}} . \tag{4.8}$$

The **moment of inertia** about an axis through the center (see Sect. 1.3.5) for expressions of Type I is

$$J = J_0 \left(1 + \frac{4}{3}\alpha\right) \tag{4.9}$$

since the value of $\langle r^2 \rangle = m_2$ as defined in (1.14) is given by

$$\langle r^2 \rangle = \langle r^2 \rangle_0 \left(1 + \frac{4}{3}\alpha\right) , \tag{4.10}$$

where

$$\langle r^2 \rangle_0 = \frac{3}{5}R_0^2 . \tag{4.11}$$

Similarly, for distributions of Type II ,

$$\langle r^2 \rangle = \langle r^2 \rangle_0 \frac{1 + \frac{5}{7}\omega}{1 + \frac{3}{5}\omega} , \tag{4.12}$$

and

$$J = J_0 \frac{1 + \frac{5}{7}\omega}{1 + \frac{3}{5}\omega} . \tag{4.13}$$

4.2 Diffuse Surface Distributions

There are a number of excellent compilations of expressions used to represent diffuse density and potential distributions such as [Collard 67], [Kim 76] and [Budzanowski 76]. The functions considered in more detail below (because of their frequency of occurrence in the literature, or general utility) fall roughly into three general categories: leptodermous, holodermous, and distributions obtained by folding.

4.2.1 Leptodermous Distributions

For leptodermous distributions (see Chap. 2) of the type often used in the macroscopic description of nuclear densities an effective sharp radius can be defined. This quantity R can be used to associate volume and surface properties with the distribution. For holodermous distributions, such as the Gaussian of Sect. 4.2.3.1 no such identification is possible.

Note: The general expression in terms of a leptodermous expansion for the Coulomb energy of a uniform distribution with a diffuse surface is,

$$E_{\text{Coul}} = E_{\text{Coul}}^0 \left(1 - \frac{5}{2}\beta^2 \cdots \right) , \tag{4.14}$$

where β is the ratio b/R defined in Eq. (2.25). For specific functional forms for the charge distribution, the Coulomb energy and the relationships between C, R and Q as defined in Chap. 2 are sometimes known to higher order (in the ratio of the diffuseness to the radius).

4.2.1.1 Fermi Function (2-parameter)

The diffuse surface distribution with half value radius C

$$\varrho(r) = \varrho_0 \left[1 + \exp\frac{r - C}{z}\right]^{-1} \tag{4.15}$$

is usually called a *Fermi function* when applied to density distributions and a *Woods-Saxon function* when applied to single particle or optical model potential wells.

So long as $e^{-C/z}$ is negligible we have (See Chap. 2)

$$b = \frac{\pi}{\sqrt{3}} z , \quad \gamma_4 = \frac{21}{5} , \quad \gamma_3 = \gamma_5 = 0 \tag{4.16}$$

and the relationships (from (2.44) and (2.45)),

$$
\begin{aligned}
C &= R\left[1 - \frac{1}{3}\left(\frac{\pi z}{R}\right)^2 + \mathcal{O}\left(\left(\frac{z}{R}\right)^6\right)\right] , \\
Q &= R\left[1 + \frac{5}{6}\left(\frac{\pi z}{R}\right)^2 - \frac{7}{24}\left(\frac{\pi z}{R}\right)^4 + \mathcal{O}\left(\left(\frac{z}{R}\right)^6\right)\right] .
\end{aligned}
\tag{4.17}
$$

The value of C can be determined from the normalization condition

$$A = \frac{4}{3}\pi R^3 \varrho_0 \tag{4.18}$$

and the (often encountered) mean square radius of the distribution $\langle r^2 \rangle$ is given by

$$\langle r^2 \rangle = \frac{3}{5}Q^2 . \tag{4.19}$$

In the application of (4.15) to the description of nuclear density distributions or potential wells the shape of the surface is sometimes modified by raising the function to various integer or noninteger powers. Expressions for calculating radial moments of powers of the Fermi function

$$I_{\nu\mu} = \int\limits_0^\infty dr \; r^\mu \left(1 + \exp\frac{r-C}{z}\right)^{-\nu} , \tag{4.20}$$

where $\mu \geq 0$ is integer and $\nu > 0$, can be written [Krivine 81], cf. also [Srivastava 82b],

$$I_{\nu\mu} = \frac{R^{\mu+1}}{\mu+1}\left[1 + (\mu+1)\sum_{k=0}^\mu \binom{\mu}{k}\eta_\nu^{(k)}\left(\frac{a}{R}\right)^{k+1} + \omega_{\nu\mu}\right] , \tag{4.21}$$

where

$$\eta_\nu^{(k)} = (-)^k \int\limits_0^\infty du \; u^k \left[\frac{1 + (-)^k e^{-u\nu}}{(1 + e^{-u})^\nu} - 1\right] \tag{4.22}$$

and $\omega_{\nu\mu}$ is of the order of $e^{-R/a}$. Recursion relations and representative values for the quantities $\eta_\nu^{(k)}$ can be found in Appendix B of [Treiner 86]. It should be noted that (4.21) is not a series expansion but (when the exponential terms can be neglected) a polynomial whose terms do not necessarily decrease with increasing k.

Because of the popularity of this function for the representation of nuclear charge distributions, the Coulomb energy has been calculated in a variety of different ways in terms of the quantities C, R and Q defined in Chap. 2.

Coulomb Energy

Direct (see Sect. 1.5.1)

In terms of E_{Coul}^0, (4.3) and β, (2.25), [Carlson 61a], [Jänecke 72a]

$$E_{\text{Coul}} = E_{\text{Coul}}^0\left(1 - \frac{5}{2}\beta^2 + 3.0216\,\beta^3 + \beta^4\cdots\right) . \tag{4.23}$$

In terms of C [Carlson 61a]

$$E_{\text{Coul}} = \frac{3}{5}\left(\frac{4}{3}\pi\varrho_0 e\right)^2 C^5 \left[1 + \frac{5}{6}\left(\frac{\pi z}{C}\right)^2 + 0.5815\left(\frac{\pi z}{C}\right)^3\right.$$
$$\left. - \frac{1}{6}\left(\frac{\pi z}{C}\right)^4 + 0.0922\left(\frac{\pi z}{C}\right)^5 + \mathcal{O}\left(e^{-z/C}\right)\right], \tag{4.24}$$

or [Lindner 68]

$$E_{\text{Coul}} = \frac{3}{5}\frac{Z^2 e^2}{C}\left[1 - \frac{7}{6}\pi^2\left(\frac{z}{C}\right)^2 + 18.031\left(\frac{z}{C}\right)^3\cdots\right] . \tag{4.25}$$

In terms of Q [Jänecke 72a]

$$E_{\text{Coul}} = \frac{3}{5}\frac{Z^2 e^2}{Q}\left[1 + 18.0295\left(\frac{z}{Q}\right)^3 - \frac{7}{8}\pi^4\left(\frac{z}{Q}\right)^4\cdots\right], \tag{4.26}$$

where it should be noted that the term quadratic in the diffuseness is missing when the energy is expressed in terms of Q. In [Jänecke 72a] the symbol R and the words *equivalent radius* are used for the quantity that we represent here by the symbol Q (2.16).

Note: To lowest order the shape dependence of the Coulomb energy does not have a diffuseness correction [Myers 66]. This is only true when the leading term is the Coulomb energy of the equivalent sharp surfaced sphere, as in (4.23) It is not true when the leading term is written in terms of C or Q as in (4.24)-(4.26) above.

Exchange (see Sect. 1.5.2)

In terms of E_{ex}^0, (1.63) and β, (2.25), as derived from [Lindner 68]

$$E_{ex} = E_{ex}^0 \left[1 - 0.736\,\beta \cdots\right] . \tag{4.27}$$

In terms of C [Lindner 68]

$$E_{ex} = -\frac{3}{4}\left(\frac{3}{2\pi}\right)^{2/3}\frac{e^2 Z^{4/3}}{C}\left[1 - 1.3355\left(\frac{z}{C}\right)\cdots\right] . \tag{4.28}$$

In terms of Q [Jänecke 72a]

$$E_{ex} = -3\left(\frac{3}{16\pi}\right)^{2/3}\frac{e^2 Z^{4/3}}{Q}\left[1 - 1.336\left(\frac{z}{Q}\right) + 7.127\left(\frac{z}{Q}\right)^2\right.$$

$$\left. - 18.210\left(\frac{z}{Q}\right)^3 + 83.406\left(\frac{z}{Q}\right)^4 \cdots\right] . \tag{4.29}$$

4.2.1.2 Fermi Function (3-parameter)

A *Generalized Fermi Function*, which includes both a central depression and diffuseness, has also been considered. One such distribution is

$$\varrho(r) = \frac{eZ}{\frac{4}{3}\pi C^3}\left[1 + \omega\left(\frac{r}{C}\right)^2\right]\left[1 + \exp\frac{r - C}{z}\right]^{-1} F^{-1} , \tag{4.30}$$

where

$$F = \left(1 + \frac{3}{5}\omega\right) + (1 + 2\omega)\,\pi^2\left(\frac{z}{C}\right)^2 \cdots . \tag{4.31}$$

Coulomb Energy [Lindner 69]

Direct (see Sect. 1.5.1)

$$E_{Coul} = \frac{3}{5}\frac{e^2 Z^2}{C F^2}\left[1 + \frac{8}{7}\omega + \frac{1}{3}\omega^2 \cdots + \left(\frac{5}{6} + 3\omega + \frac{3}{2}\omega^2\right)\pi^2\left(\frac{z}{C}\right)^2\right.$$

$$\left. + 18.031\,(1 + \omega)^2\left(\frac{z}{C}\right)^3 \cdots\right] . \tag{4.32}$$

Exchange (see Sect. 1.5.2)

$$E_{\text{ex}} = -\frac{3}{4}\left(\frac{3}{2\pi}\right)^{2/3}\frac{e^2}{C}\left(\frac{Z}{F}\right)^{4/3}$$

$$\times\left[1 + \frac{4}{5}\omega + \frac{2}{21}\omega^2 - 1.3355\,(1+\omega)^{4/3}\,\frac{z}{C}\cdots\right]. \tag{4.33}$$

The expression relating R and C for such distributions [Myers 69], which is

$$C = R\left[1 - \beta^2\frac{1+2\omega}{1+\omega}\cdots\right], \tag{4.34}$$

can be used to convert (4.32) to an expression in terms of E^0_{Coul} (4.3) and β (2.25).

$$E_{\text{Coul}} = E^0_{\text{Coul}}\left\{\frac{1 + \frac{8}{7}\omega + \frac{1}{3}\omega^2}{\left(1 + \frac{3}{5}\omega\right)^2}\right.$$

$$\left. - \frac{5}{2}\beta^2\frac{1+\omega}{1+\frac{3}{5}\omega}\left[1 + \frac{8}{175}\frac{\omega(1+2\omega)(1+\frac{1}{3}\omega)}{(1+\omega)^2\,(1+\frac{3}{5}\omega)^2}\right]\right\}. \tag{4.35}$$

If the central depression is based on the equivalent sharp radius R rather than C, i.e.

$$\varrho(r) \propto \left[1 + \omega\left(\frac{r}{R}\right)^2\right]\left[1 + \exp\left(\frac{r-C}{z}\right)\right]^{-1}, \tag{4.36}$$

then the expression for the Coulomb energy is [Myers 69]

$$E_{\text{Coul}} = E^0_{\text{Coul}}\left[\frac{1 + \frac{8}{7}\omega + \frac{1}{3}\omega^2}{\left(1 + \frac{3}{5}\omega\right)^2} - \frac{5}{2}\beta^2\frac{1+\omega}{1+\frac{3}{5}\omega}\cdots\right]. \tag{4.37}$$

An alternative to the *Generalized Fermi Function* of Eq. (4.30), suggested in [Ford 69] is

$$\varrho = \varrho_0\left\{1 + \exp[-(r-C)/a]\right\}^{-1}\cdot\begin{cases} 1 + \omega(r/C)^2 & , r \leq C \\ 1 + \omega & , r > C. \end{cases} \tag{4.38}$$

Another alternative, suggested by (4.4), is

$$\varrho = \varrho_0\left\{1 + \alpha\left[-\frac{3}{5} + \left(\frac{r}{C}\right)^2\right]\right\}\left\{1 + \exp\left(-\frac{r-C}{a}\right)\right\}^{-1}. \tag{4.39}$$

Note: The quantity ϱ_0 means something slightly different in each of the expressions (4.30), (4.38) and (4.39).

4.2.1.3 Symmetrized Fermi Distribution

In [Münchow 79] and [Grammaticos 82] a distribution is employed which is identical to the Fermi function (4.15) for large values of the radius parameter

C. This function

$$\varrho(r) = \frac{\sinh(C/a)}{\cosh(C/a) + \cosh(r/a)} , \qquad (4.40)$$

has a number of nice properties that are missing for Fermi functions. One such property is that the derivative of the function at the origin vanishes, $\varrho'(0) = 0$. Another feature is that the exact expression for the normalization of the density is

$$A = \frac{4}{3}\pi C^3 \varrho_0 \left[1 + \left(\frac{\pi a}{C}\right)^2 \right] , \qquad (4.41)$$

which shows that $A = 0$ when $C = 0$ (this is not true for a Fermi function). In addition there is an exact expression for the mean square radius $\langle r^2 \rangle$, which is

$$\langle r^2 \rangle = \frac{3}{5}C^2 \left[1 + \frac{7}{3}\left(\frac{\pi a}{C}\right)^2 \right] . \qquad (4.42)$$

As with the Fermi function itself, the relations between C, R and Q are given by (4.17). A host of other useful properties of this function can be found in [Grammaticos 82].

Radial moments of powers of the function (4.40) are identical to those of the Fermi function except that the exponential terms are missing. For (4.40) the polynomial terms in (4.21) constitute the *exact* result.

4.2.1.4 Trapezoidal Distribution

For the trapezoidal distribution defined by

$$\varrho(r) = \frac{1}{2}\varrho_0 \left[1 - \frac{n}{\max(a, |n|)} \right] , \qquad (4.43)$$

where

$$r = C + n \qquad (4.44)$$

and

$$b = \frac{a}{\sqrt{3}} , \quad \gamma_4 = \frac{9}{5} , \quad \gamma_3 = \gamma_5 = 0 . \qquad (4.45)$$

This distribution is discussed in [Süssmann 75] and the relationships between C, R and Q are given in Chap. 2. The direct part of the Coulomb energy is given by [Gunter 59]

$$E_{\text{Coul}} = \frac{8\pi^2}{3}\varrho_0^2 e^2 \left[\frac{2}{5}C^5 + \frac{1}{3}C^3 a^2 + \frac{1}{5}C^2 a^3 + \frac{1}{15}Ca^4 - \frac{1}{105}a^5 \right] . \qquad (4.46)$$

4.2.2 Hill-Ford Distributions

In their articles on nuclear charge distributions Hill and Ford [Hill 54], [Ford 54] consider a number of different parameterizations for the nuclear charge distributions that can be varied from *leptodermous* (Sect. 4.2.1) to *holodermous* (Sect. 4.2.3). They formulate their expressions in a dimensionless way with the definitions

35

$$\varrho(r) \quad = \varrho_0\, f_\lambda(x)$$
$$x \quad := r/a\,. \tag{4.47}$$

Here ϱ_0 is the central charge density, $f_\lambda(0) = 1$; λ represents any number of parameters used to define the shape of the distribution, and a the range parameter, determines the radial extent of the distribution. Two functionals of f are defined by

$$I_f(x) \quad = \quad \int_0^x f(x)\, x^2 \mathrm{d}x\,; \tag{4.48}$$

$$J_f(x) \quad = \quad [I_f(\infty)]^{-1} \int_0^x I_f(x)\, x^{-2} \mathrm{d}x\,. \tag{4.49}$$

Then the normalizing condition is

$$4\pi\, r_0^3\, \varrho_0\, I_f(\infty) = Ze\,, \tag{4.50}$$

and the electrostatic potential is

$$V = -\frac{Ze}{a}\, J_f(x)\,. \tag{4.51}$$

Family I

$$f_n(x) = \frac{1}{n!} \int_x^\infty x^n e^{-x} \mathrm{d}x = \sum_{k=0}^n \frac{x^k}{k!} e^{-x} \quad n = 0,1,2\cdots\,.$$

$$I_f(\infty) = \frac{1}{3}(n+1)(n+2)(n+3),$$

$$J_f(x) = \frac{1}{x}(1 - e^{-x}) - e^{-x} \sum_{k=0}^n b_k \frac{x^k}{k!}\,,$$

$$b_k = \frac{n+1}{n+2} - \frac{k}{k+1} - \frac{n(n+1) - (k-1)k}{2(n+1)(n+2)(n+3)}\,. \tag{4.52}$$

For $n = 0$, this is an exponential, for $n = 1$, a modified exponential. As $n \to \infty$, f approaches a square distribution, but the high-n members of the family are not feasible for calculation.

Family IIa

$$f_n(x) = \frac{1}{1 - \frac{1}{2}e^{-n}} \cdot \left\{ \begin{array}{ll} 1 - \frac{1}{2}e^{-n}e^x & ,\ x \le n \\[2mm] \frac{1}{2}e^n e^{-x} & ,\ x \ge n \end{array} \right\},\quad 0 \le n \le 1$$

$$I_f(\infty) = \frac{e^{-n} + 2n + \frac{1}{3}n^3}{1 - \frac{1}{2}e^{-n}}$$

$$J_f(x) = \frac{1 + \frac{1}{2}n^2 - \frac{1}{6}x^2 + e^{-n}\left(\frac{1 - e^x}{x} + \frac{1}{2}e^x\right)}{e^{-n} + 2n + \frac{1}{3}n^3}\,,\quad x < n$$

$$= \frac{1}{x} - \frac{e^{n-x}\left(\frac{1}{x} + \frac{1}{2}\right)}{e^{-n} + 2n + \frac{1}{3}n^3}\,,\quad x > n\,. \tag{4.53}$$

As n varies from 0 to 1, f varies from an exponential toward a shape roughly Gaussian in appearance.

In problem (3-9) of [Preston 62] it is noted that if $x = nr/c$ then the Coulomb energy of such a distribution is, to a good approximation, given by

$$E_{\text{Coul}} = \frac{3}{5}\frac{Z^2 e^2}{c}(1 + 6n^{-2})^{-1}\left\{1 - \frac{n^{-2}}{1 + 6n^{-2}}\left(1 - \frac{75}{8n} + \frac{15}{n^2} - \frac{315}{16n^3}\right)\right\}. \quad (4.54)$$

Family IIb

$$f_n(x) = \frac{1}{1 - \frac{1}{2}e^{-n}} \times \left\{\begin{array}{ll} 1 - \frac{1}{2}e^{-n(1-x)} & , \ x \leq 1 \\[2mm] \frac{1}{2}e^{-n(x-1)} & , \ x \geq 1 \end{array}\right\}, \quad 0 \leq n \leq \infty$$

$$I_f(\infty) = \frac{e^{-n} + 2n + \frac{1}{3}n^3}{n^3(1 - \frac{1}{2}e^{-n})}$$

$$J_f(x) = \frac{\frac{1}{n^2} + \frac{1}{2} - \frac{1}{6}x^2\frac{e^{-n}}{n^2}\left(\frac{1 - e^{nx}}{nx} + \frac{1}{2}e^{nx}\right)}{\frac{1}{3} + \frac{2}{n^2} + \frac{e^{-n}}{n^3}}, \quad x < 1$$

$$= \frac{1}{x} - e^{n(1-x)}\frac{\frac{1}{x} + \frac{n}{2}}{e^{-n} + 2n + \frac{1}{3}n^3}, \quad x > 1. \quad (4.55)$$

As n varies from 1 to infinity, f varies from a shape roughly Gaussian in appearance to a square shape (uniform distribution). Families IIa and IIb form a singly connected family (being identical at $n = 1$), but are distinguished in order that the range parameter a may have a close relation to the size. For family IIa, the relevant distance is the decay length for $x > n$. For family IIb, the relevant distance is the interval out to the point where the distribution begins to fall exponentially.

Family III

$$f_{ns}(x) = \frac{(\sinh sx)/sx}{1 - \frac{1}{2}e^{-n}} \times \left\{\begin{array}{ll} 1 - \frac{1}{2}e^{-n(1-x)} & , \ x \leq 1 \\[2mm] \frac{1}{2}e^{-n(x-1)} & , \ x \geq 1 \end{array}\right\}, \quad n \geq 1, \ n > s.$$

$$I_f(\infty) = \frac{1}{1 - \frac{1}{2}e^{-n}}\left[\frac{s\cosh s - \sinh s}{s^3}\right.$$

$$\left. + \frac{(n^2 - s^2)s\cosh s + (n^2 + s^2)\sinh s}{s(n^2 - s^2)^2} + \frac{ne^{-n}}{(n^2 - s^2)^2}\right]$$

$$J_f(x) = \frac{1}{D}\left\{\frac{n^2}{s^2(n^2 - s^2)}\cosh s + \frac{ne^{-n}}{(n^2 - s^2)^2}\frac{1}{x} - \frac{\sinh sx}{s^3 x}\right.$$

$$\left. + \frac{e^{-n}}{2s(n^2 - s^2)^2}\frac{e^{nx}}{x}[(n^2 + s^2)\sinh sx - 2ns\cosh sx]\right\}, \quad x < 1$$

$$= \frac{1}{x} - \frac{e^n}{2(n^2 - s^2)^2 s}\frac{e^{-nx}}{x}$$

$$\times \frac{1}{D}[2ns\cosh sx + (n^2 + s^2)\sinh sx], \quad x > 1$$

(4.56)

$$D = \left(1 - \frac{1}{2}e^{-n}\right)I_f(\infty).$$

37

This family is a generalization of family IIb and includes shapes peaked at the edge instead of at the center.

Family IV

This family is made of various separate simple distributions, some of which are special cases of the previous three families.

A. Exponential:
$$f = e^{-x}$$
$$I_f(\infty) = 2$$
$$J_f(x) = \frac{1}{x} - e^{-x}\left(\frac{1}{x} + \frac{1}{2}\right). \tag{4.57}$$

B. Modified exponential:
$$f = (1+x)e^{-x}$$
$$I_f(\infty) = 8$$
$$J_f(x) = \frac{1}{x} - e^{-x}\left[\frac{1}{x} + \frac{5}{8} + \frac{x}{8}\right]. \tag{4.58}$$

C. Gaussian:
$$f = e^{-x^2}$$
$$I_f(\infty) = \tfrac{1}{4}\sqrt{\pi}$$
$$J_f(x) = \frac{1}{x}\frac{2}{\sqrt{\pi}}\int_0^x e^{-z^2}\mathrm{d}z = \frac{\mathrm{erf}\,x}{x}. \tag{4.59}$$

D. Modified Gaussian:
$$f = (1+x^2)e^{-x^2}$$
$$I_f(\infty) = \tfrac{5}{8}\sqrt{\pi}$$
$$J_f(x) = \frac{\mathrm{erf}\,x}{x} - \frac{2}{5\sqrt{\pi}}e^{-x^2}. \tag{4.60}$$

4.2.3 Holodermous Distributions

As pointed out in [Myers 76], "The word *leptodermous*, like a number of other words used in nuclear physics, (such as nucleus or fission) has been taken from the field of biology. It is an adjective meaning *having a thin skin*. One difference is that we use the word here in the relative rather than the absolute sense. Consequently, even elephants (called pachyderms because of their thick skins) are leptodermous in our usage. Perhaps an appropriate antonym would be *holodermous*, meaning *all skin*, a term that would apply, for example, to the lightest nuclei or to the electron distribution in atoms."

4.2.3.1 Gaussian

For a Gaussian distribution (*generalized* by the addition of a term quadratic in r/a_g) the density is given by [Carlson 61a]

$$\varrho(r) = \varrho_0 \left[1 + \omega \left(\frac{r}{a_g} \right)^2 \right] \exp \left[- \left(\frac{r}{a_g} \right)^2 \right] \tag{4.61}$$

and the particle number Z by

$$Z = \pi^{3/2} \varrho_0 a_g^3 \left(1 + \frac{3}{2} \omega \right) , \tag{4.62}$$

and the direct Coulomb energy by

$$E_{\text{Coul}} = \sqrt{\frac{\pi^5}{2}} \varrho_0^2 e^2 a_g^5 \left(1 + \frac{5}{2} \omega + \frac{27}{16} \omega^2 \right) . \tag{4.63}$$

(See also Family IV in Sect. 4.2.2).

4.2.3.2 Exponential

Two forms of this distribution are considered under Family IV in Sect. 4.2.2.

4.2.4 Folded Distributions

The general discussion of distributions generated by folding is the subject of Chap. 3. Here we consider only the case of Yukawa (3.11) and Gaussian (3.14) folding functions and a spherical generating function with a sharp surface

$$g(r) = \varrho_0 \, \Theta(R - r) . \tag{4.64}$$

The analytic expressions for the spherically symmetric distributions that result are given below. When $R \gg b_f$ (see (2.49)) the special relationships of Chap. 2 apply relating the geometrical quantities C, R, and Q.

4.2.4.1 Yukawa Folding Function

For the Yukawa folding function, (3.11), convolution with (4.64) above yields [Bolsterli 72]

$$\varrho(r) = \begin{cases} \varrho_0 \left[1 - \left(1 + \dfrac{R}{a} \right) \exp(-R/a) \, \dfrac{\sinh(r/a)}{r/a} \right] & , \; r \leq R \\[4mm] \varrho_0 \left[\dfrac{R}{a} \cosh(R/a) - \sinh(R/a) \right] \dfrac{\exp(-r/a)}{r/a} & , \; r \geq R . \end{cases} \tag{4.65}$$

When $R \gg a$ the width of the surface b (2.49) of such a distribution is simply the width of the generating function (3.13),

$$b_{\text{Y}} = a\sqrt{2} . \tag{4.66}$$

The higher order surface moment γ_4 from Sect. 2.4 is

$$\gamma_4 = 6 \tag{4.67}$$

and all the odd γ's are zero.

The Coulomb energy of such a distribution is [Krappe 76]

$$E_{\text{Coul}} = E^0_{\text{Coul}} \left[1 - 5\sigma^2 + \frac{75}{8}\sigma^3 - \frac{105}{8}\sigma^5 \right.$$
$$\left. + \frac{15}{8}\sigma^2(1+\sigma)\left(2 + 7\sigma + 7\sigma^2\right) e^{-2/\sigma} \right] , \qquad (4.68)$$

where $\sigma = a/R$.

4.2.4.2 Gaussian Folding Function

For the Gaussian folding function, (3.14) convolution with (4.64) yields [Krappe 76]

$$\varrho(r) = \frac{\varrho_0}{2\sqrt{\pi}(r/a)} \left(e^{-(r/a+R/a)^2} - e^{-(r/a-R/a)^2} \right)$$
$$+ \frac{\varrho_0}{2}(\text{erf}\,(R/a - r/a) + \text{erf}\,(R/a + r/a)) . \qquad (4.69)$$

As in the previous section the surface width b (2.49) of this distribution (for $R \gg a$) is simply that of the folding function (3.16)

$$b_G = a/\sqrt{2} . \qquad (4.70)$$

The higher order surface moment γ_4 from Sect. 2.4 is

$$\gamma_4 = 3 , \qquad (4.71)$$

and all the odd γ's are zero.

Chapter 5

Spheroidal Deformations

5.1 Spheroids

Spheroidal deformations are useful in describing the ground state deformations of many nuclei and are also suited for describing larger prolate and oblate deformations in schematical calculations. Here and in the following, (ϱ, z) are axially symmetric cylindrical coordinates, $\varrho_s \equiv \varrho_s(z)$ is the surface function, z_0 is half the length, and $\varrho_n = \varrho_s(0)$ is the neck radius for symmetric shapes.

Definition

$$\frac{\varrho_s^2}{a^2} + \frac{z^2}{c^2} = 1 \,, \tag{5.1}$$

where

$$\left.\begin{array}{rclcl} a & = & \varrho_n & = & \mathrm{minor(major)} \\ c & = & z_0 & = & \mathrm{major(minor)} \end{array}\right\} \text{semiaxis for prolate(oblate) spheroid.} \tag{5.2}$$

Volume conservation

$$a^2 c = R_0^3 \,. \tag{5.3}$$

If the eccentricity e is defined by

$$e^2 = \begin{cases} 1 - \dfrac{a^2}{c^2} & , \text{ prolate} \\[2mm] \dfrac{a^2}{c^2} - 1 & , \text{ oblate,} \end{cases} \tag{5.4}$$

then for prolate ellipsoids

$$\begin{array}{rcl} a & = & R_0(1 - e^2)^{1/6} \\ c & = & R_0(1 - e^2)^{-1/3} \,. \end{array} \tag{5.5}$$

If we define

$$\zeta_0 = c/R_0 \tag{5.6}$$

as the major (minor) semiaxis in units of R_0 then

$$\varrho_n = \zeta_0^{-1/2} = a/R_0 \tag{5.7}$$

is the minor (major) semiaxis or neck radius in units of R_0. Furthermore, in the natural units of Chap. 1 the relative mean square radius, quadrupole moment and moments of inertia are given by

$$r^2_{\text{rms}} = \frac{2}{5\zeta_0} + \frac{\zeta_0^2}{5}$$

$$Q = \frac{8}{15}\pi\,\zeta_0^2 e^2 \tag{5.8}$$

$$\mathcal{J}_\parallel = \zeta_0^{-1}$$

$$\mathcal{J}_\perp = \frac{1}{2}(\zeta_0^2 + \zeta_0^{-1})$$

$$\mathcal{J}_{\text{eff}}^{-1} = \zeta_0 - \frac{2}{\zeta_0^2 + \zeta_0^{-1}}\,. \tag{5.9}$$

According to [Carlson 61b] the angular mean square radius (1.32) simply reads, cf. Eq. (5.11),

$$\langle r^2 \rangle = B_{\text{Coul}}. \tag{5.10}$$

The relative energies read, cf. [Beringer 61]

$$B_{\text{Coul}} = \frac{(1-e^2)^{1/3}}{2e}\,\log\frac{1+e}{1-e} \qquad\qquad\text{, prolate}$$

$$= \frac{(1+e^2)^{1/3}}{e}\,\arctan e \qquad\qquad\text{, oblate}$$

$$B_{\text{surf}} = \frac{(1-e^2)^{1/3}}{2}\left(1 + \frac{\arcsin e}{e\sqrt{1-e^2}}\right) \qquad\text{, prolate}$$

$$= \frac{(1+e^2)^{1/3}}{2}\left(1 + \frac{\log(e+\sqrt{1+e^2})}{e+\sqrt{1+e^2}}\right) \quad\text{, oblate} \tag{5.11}$$

$$B_{\text{curv}} = \frac{1}{2(1-e^2)^{1/3}} + \frac{(1-e^2)^{2/3}}{4e}\,\log\frac{1+e}{1-e} \quad\text{, prolate}$$

$$= \frac{1}{2(1+e^2)^{1/3}} + \frac{(1+e^2)^{2/3}}{2e}\,\arctan e \quad\text{, oblate}$$

$$B_{\text{comp}} = B_{\text{surf}}^2\,.$$

With the abbreviations

$$K = \frac{1}{2e}\ln\frac{1+e}{1-e}$$

$$A = \zeta_0^{1/2}\,\frac{\arcsin e}{e}$$

$$L = K + \frac{K-1}{e^2}$$

$$M = \frac{1}{e^2}\left(K - 3\,\frac{K-1}{e^2}\right)$$

$$\overline{\Phi}_s = \frac{3}{32}\zeta_0^{-1}[8LB_{\text{surf}} + M(A - \zeta_0^{-1}(1-2e^2))]$$

$$\overline{\Phi_s^2} = \frac{9}{32}\zeta_0^{-2}\left[2L^2 B_{\text{surf}} + \frac{1}{8}M\Big\{A(4L+M)\right.$$
$$\left. - \zeta_0^{-1}\Big(4L(1-2e^2) + M(1+\frac{2}{3}e^2 - \frac{8}{3}e^4)\Big)\Big\}\right] \tag{5.12}$$

the redistribution energies become

$$B_{\text{red}} = \frac{3}{4}\zeta_0^{-2}\left(3K^2 - \frac{10}{e^2}K(K-1) + \frac{15}{e^4}(K-1)^2\right)$$

$$B_{\text{sr1}} = (5\overline{\Phi_s})^2 - 60\overline{\Phi_s}B_{\text{Coul}}B_{\text{surf}} + (6B_{\text{Coul}}B_{\text{surf}})^2$$

$$B_{\text{sr2}} = 25\overline{\Phi_s^2} - 60\overline{\Phi_s}B_{\text{Coul}} + 36B_{\text{Coul}}^2 B_{\text{surf}} . \tag{5.13}$$

The relative Coulomb potential and Coulomb potential at the surface read

$$\Phi(\varrho,\zeta) = \frac{3}{4}\zeta_0^{-3}\left[2\zeta_0^2 K - \frac{2}{e^2}(K-1)^2\zeta^2 - \frac{1}{e^2}(\zeta_0^3 - K)\varrho^2\right]$$

$$\Phi_s(\zeta) = \frac{3}{4}\zeta_0^{-3}\left[\zeta_0^2 L + e^2 M\zeta^2\right] . \tag{5.14}$$

The rotational energy becomes [Beringer 61] (note that moments of inertia are measured in units of the rigid body moment of inertia of a sphere)

$$B_{\text{rot}} = \mathcal{J}_\perp^{-1} = \frac{2(1-e^2)^{2/3}}{2-e^2} \quad , \text{ prolate} \tag{5.15}$$

$$= \mathcal{J}_\parallel^{-1} = (1+e^2)^{-1/3} \quad , \text{ oblate.}$$

The hydrodynamic mass parameter and the viscosity and one body dissipation coefficients of Sect. 1.8 with respect to the elongation $\zeta_0 = c/R_0$ read

$$B_{\zeta_0} = \frac{1}{5}\left(1 + \frac{1}{2\zeta_0^3}\right)$$

$$Z_{\zeta_0} = \frac{1}{\zeta_0^2}$$

$$D_{\zeta_0} = \frac{3}{32e^4\zeta_0^{5/2}}\left[\left(9 - 8e^2 + \frac{8}{3}e^4\right)\frac{\arcsin e}{e} - (9 - 2e^2)\sqrt{1-e^2}\right] , \tag{5.16}$$

where at sphericity, $D_{\zeta_0}(\zeta_0 = 1) = 1/5$.

5.2 Nilsson Potential

In Nilsson's notation [Nilsson 55]: If

$$V_{\text{osc}}(\varrho, z) = \frac{1}{2}M(\omega_\perp^2 \varrho^2 + \omega_z^2 z^2) \tag{5.17}$$

is the spheroidal oscillator potential, then equipotential surfaces labelled by the constant $\overset{\circ}{\omega}_0$ are given by

$$\varrho_s^2 \omega_\perp^2 + z^2 \omega_z^2 = R_0^2 \overset{\circ}{\omega}_0^2 \tag{5.18}$$

and the half axes of the equipotential surfaces read

$$a = R_0 \overset{\circ}{\omega}_0 / \omega_\perp$$
$$c = R_0 \overset{\circ}{\omega}_0 / \omega_z . \tag{5.19}$$

The deformation parameters ω_z, $\omega_\perp$ are replaced by the two parameters ε, $\omega_0(\varepsilon)$ defined by

$$\omega_z = \omega_0(\varepsilon)\left(1 - \frac{2}{3}\varepsilon\right)$$
$$\omega_\perp = \omega_0(\varepsilon)\left(1 + \frac{1}{3}\varepsilon\right) , \tag{5.20}$$

which, by means of the volume conservation

$$\omega_\perp^2 \omega_z = \overset{\circ}{\omega}_0^3 \tag{5.21}$$

convert to the semiaxes a, c by means of

$$\varepsilon = 3\,\frac{c-a}{2c+a}$$

$$\omega_0(\varepsilon) = \overset{\circ}{\omega}_0 \left(1 - \frac{1}{3}\varepsilon^2 - \frac{2}{27}\varepsilon^3\right)^{-1/3} \tag{5.22}$$

or

$$\frac{a}{c} = \frac{1 - \frac{2}{3}\varepsilon}{1 + \frac{1}{3}\varepsilon} . \tag{5.23}$$

From Eq. (5.23) one derives the following limits for ε :

$$-3 < \varepsilon < 3/2 . \tag{5.24}$$

Other notations employed by [Schütte 75]

$$\alpha_S = \frac{\overset{\circ}{\omega}_0}{\omega_z} = \left(\frac{1 + \frac{1}{3}\varepsilon}{1 - \frac{2}{3}\varepsilon}\right)^{2/3} \tag{5.25}$$

and by [Wilkins 76]

$$\frac{a}{c} = \frac{1 - \frac{1}{3}\beta_W}{1 + \frac{2}{3}\beta_W}. \tag{5.26}$$

The relative quadrupole moment in units of $3AR_0^2/4\pi$ [Sobiczewski 69] and the hydrodynamic mass parameter in units of MR_0^2 [Nilsson 69], [Möller 81] are given by

$$B_\varepsilon = \frac{2}{15}\,\frac{1 + \frac{2}{9}\varepsilon^2}{(1 - \frac{2}{3}\varepsilon)^2}\left(1 - \frac{1}{3}\varepsilon^2 - \frac{2}{27}\varepsilon^3\right)^{-4/3}$$

$$Q_\varepsilon = \frac{8}{15\pi}\left(1 - \frac{1}{3}\varepsilon^2 - \frac{2}{27}\varepsilon^3\right)^{2/3}\left[\left(1 - \frac{2}{3}\varepsilon\right)^{-2} - \left(1 + \frac{1}{3}\varepsilon\right)^{-2}\right]. \tag{5.27}$$

For convenience, Nilsson [Nilsson 55] abbreviates

$$(\eta\kappa) := \varepsilon \frac{\omega_0(\varepsilon)}{\overset{\circ}{\omega}_0} = \varepsilon \left(1 - \frac{1}{3}\varepsilon^2 - \frac{2}{27}\varepsilon^3\right)^{-1/3} . \qquad (5.28)$$

Transforming into the so called stretched cylindrical coordinates (ϱ_t , z_t)

$$\varrho_t = \varrho\sqrt{M\omega_\perp/\hbar}$$
$$z_t = z\sqrt{M\omega_z/\hbar} \qquad (5.29)$$

and the associated stretched polar coordinates (r_t , θ_t)

$$r_t^2 = \varrho_t^2 + z_t^2$$
$$\cos\theta_t = z_t/r_t$$

$$= \sqrt{\frac{1 - \frac{2}{3}\varepsilon}{1 + \frac{1}{3}\varepsilon - \varepsilon\cos^2\theta}} \cos\theta , \qquad (5.30)$$

or

$$r_t^2 = \frac{M}{\hbar}\omega_0(\varepsilon) \left(\varrho^2 \left(1 + \frac{1}{3}\varepsilon\right) + z^2 \left(1 - \frac{2}{3}\varepsilon\right)\right)$$
$$\frac{1}{\cos^2\theta_t} - 1 = \left(\frac{1}{\cos^2\theta} - 1\right)\frac{1 + \frac{1}{3}\varepsilon}{1 - \frac{2}{3}\varepsilon} , \qquad (5.31)$$

the potential assumes a simple form

$$V_{\text{osc}}(\varrho_t, z_t) = \frac{\hbar}{2}(\varrho_t^2\omega_\perp + z_t^2\omega_t)$$
$$= \frac{\hbar}{2}\omega_0(\varepsilon)\, r_t^2 \left(1 - \frac{2}{3}\varepsilon P_2(\cos\theta_t)\right) , \qquad (5.32)$$

where P_2 is the Legendre polynomial of order two.

An equivalent description is provided by the δ-parameter [Nilsson 55], [Bohr 75]

$$\omega_z^2 = \omega_0^2(\delta)\left(1 - \frac{4}{3}\delta\right)$$
$$\omega_\perp^2 = \omega_0^2(\delta)\left(1 + \frac{2}{3}\delta\right) , \qquad (5.33)$$

which can be converted using volume conservation, Eq. (5.21), to the semiaxes by means of

$$\delta = \frac{3}{2}\frac{c^2 - a^2}{2c^2 + a^2}$$

$$\omega_0(\delta) = \overset{\circ}{\omega}_0 \left(1 - \frac{4}{3}\delta^2 - \frac{16}{27}\delta^3\right)^{-1/6}$$

or

$$\frac{a}{c} = \sqrt{\frac{1 - \frac{4}{3}\delta}{1 + \frac{2}{3}\delta}} . \qquad (5.34)$$

45

Chapter 6

Small Deformations

Formulae in this chapter refer to series expansions and, hence, are only valid approximately, if convergent, unless otherwise noted. Here and in the following, (r, θ, φ) are spherical coordinates, $R(\theta)$ or $R(\theta, \varphi)$ are the surface functions, $P_l(\cos\theta)$ is the Legendre polynomial and $Y_{lm}(\theta, \varphi)$ are the spherical harmonics.

6.1 Spheroidal Expansion

Slightly spheroidally distorted spheres are obtained in letting the eccentricity (Eq. 5.4) be a small expansion parameter. Then for prolate shapes

$$a = \varrho_n = R_0\left(1 - \frac{1}{6}e^2 - \frac{5}{72}e^4\cdots\right)$$

$$c = z_0 = R_0\left(1 + \frac{1}{3}e^2 + \frac{2}{9}e^4\cdots\right) \tag{6.1}$$

For oblate spheroids, e^2 is to be replaced by $-e^2$. Up to order e^8 the expansions of the geometrical quantities are given, in the natural units discussed in Chap. 1, by the expressions

$$r_{\text{rms}}^2 = \frac{3}{5}\left(1 - \frac{1}{9}e^2 + \frac{1}{9}e^4 + \frac{10}{81}e^6 + \frac{10}{81}e^8\cdots\right)$$

$$\varrho_n = 1 - \frac{1}{6}e^2 - \frac{5}{72}e^4 - \frac{55}{1296}e^6 - \frac{935}{31104}e^8\cdots$$

$$\zeta_0 = 1 + \frac{1}{3}e^2 + \frac{2}{9}e^4 + \frac{14}{81}e^6 + \frac{35}{243}e^8\cdots$$

$$Q = \frac{8}{15}\pi e^2\left(1 + \frac{2}{3}e^2 + \frac{5}{9}e^4 + \frac{40}{81}e^6\cdots\right) \tag{6.2}$$

$$\mathcal{J}_\parallel = 1 - \frac{1}{3}e^2 - \frac{1}{9}e^4 - \frac{5}{81}e^6 - \frac{10}{243}e^8\cdots$$

$$\mathcal{J}_\perp = 1 + \frac{1}{6}e^2 + \frac{2}{9}e^4 + \frac{35}{162}e^6 + \frac{50}{243}e^8\cdots$$

$$\mathcal{J}_{\text{eff}} = \frac{1}{2}e^2 + \frac{5}{12}e^4 + \frac{85}{216}e^6 + \frac{71}{324}e^8\cdots \tag{6.3}$$

$$B_{\text{Coul}} = 1 - \frac{1}{45}e^4 - \frac{64}{2835}e^6 - \frac{58}{2835}e^8\cdots$$

$$B_{\text{surf}} = 1 + \frac{2}{45}e^4 + \frac{116}{2835}e^6 + \frac{101}{2835}e^8\cdots$$

$$B_{\text{curv}} = 1 + \frac{2}{45}e^4 + \frac{136}{2835}e^6 + \frac{131}{2835}e^8\cdots$$

$$B_{\text{comp}} = 1 + \frac{4}{45}e^4 + \frac{232}{2835}e^6 + \frac{346}{4725}e^8 \cdots$$

$$B_{\text{red}} = 1 + \frac{2}{45}e^4 + \frac{136}{2835}e^6 + \frac{4009}{99225}e^8 \cdots$$

$$B_{\text{sr1}} = 1 - \frac{2}{45}e^4 - \frac{124}{2835}e^6 - \frac{1219}{33075}e^8 \cdots$$

$$B_{\text{sr2}} = 1 + 0e^4 + 0e^6 + \frac{16}{11025}e^8 \cdots \tag{6.4}$$

Since the expansions of $(B_{\text{surf}} - 1)$ and $(B_{\text{Coul}} - 1)$ start with e^4- terms, the stiffness and hydrodynamic mass parameters of Sect. 1.8 only exist if e^2 is taken as the expansion parameter about the spherical shape. For a nucleus with fissility x,

$$C_{e^2} = \frac{4}{45}(1 - x)$$

$$B_{e^2} = \frac{1}{30}$$

$$\omega_{e^2}^2 = \frac{8}{3}(1 - x). \tag{6.5}$$

6.2 Harmonic Expansions

6.2.1 α_n-Parameterization

The basic definition of shape [Present 40], [Present 46], [Hill 53], [Swiatecki 56a], [Nix 67] is given by:

$$R(\theta) = \lambda^{-1}R_0 \left(1 + \sum_{n=1}^{N} \alpha_n P_n(\cos\theta)\right). \tag{6.6}$$

Here N is a cutoff parameter, λ is the volume conservation,

$$\lambda^3 = 1 + \sum_{n=1}^{N} \frac{3}{2n+1}\alpha_n^2 + \frac{1}{2}\sum_{l,m,n=1}^{N}(lmn)\alpha_l\alpha_m\alpha_n, \tag{6.7}$$

where the n-symbol-bracket is defined by

$$(k_1 k_2 \cdots k_n) = \int_{-1}^{+1} P_{k_1}(\mu) P_{k_2}(\mu) \cdots P_{k_n}(\mu) \, d\mu. \tag{6.8}$$

One-,two and three-symbol brackets can be written in closed form,

$$(i) = 2\delta_{i0} \tag{6.9}$$

$$(ij) = \frac{1}{2i+1}\delta_{ij} \tag{6.10}$$

$$(ijk) = \frac{2(i+j-k)!(j+k-i)!(k+i-j)!\,p!^2}{(2p+1)!(p-i)!^2(p-j)!^2(p-k)!^2} \tag{6.11}$$

if $2p = i + j + k$ is an even number and if one argument does not exceed the sum of all others, otherwise it vanishes. Table 6.1 gives a collection of three- and four- symbol brackets. Center-of-mass conservation in case of asymmetric shapes is achieved by eliminating α_1, [Present 40],

$$
\begin{aligned}
\alpha_1 &= -9 \sum_{n=2}^{N} \frac{n+1}{(2n+1)(2n+3)} \alpha_n \alpha_{n+1} + \cdots \\
&= -\frac{27}{35} \alpha_2 \alpha_3 \cdots .
\end{aligned}
\tag{6.12}
$$

If only $n = 2$ is retained volume conservation is given exactly by

$$
\lambda^3 = 1 + \frac{3}{5}\alpha_2^2 + \frac{2}{35}\alpha_2^3
\tag{6.13}
$$

and if only $n = 2, 4$ are retained the scission configuration of tangent fragments is given by the line

$$
4\alpha_2 - 3\alpha_4 = 8 .
\tag{6.14}
$$

In this case, Fig. 6.1 displays the shapes contained in this parameterization. Dimensionless neck radius and half length read

$$
\varrho_n = \lambda^{-1} \left(1 - \frac{1}{2}\alpha_2 + \frac{3}{8}\alpha_4 - \frac{5}{16}\alpha_6 \cdots \right)
\tag{6.15}
$$

$$
\zeta_0 = \lambda^{-1} \sum_{n=1}^{N} \alpha_n .
\tag{6.16}
$$

The quantity α_2 is sometimes simply denoted by α. For N=5 and using the natural units of Chap. 1 the explicit expansions become (for the multipole

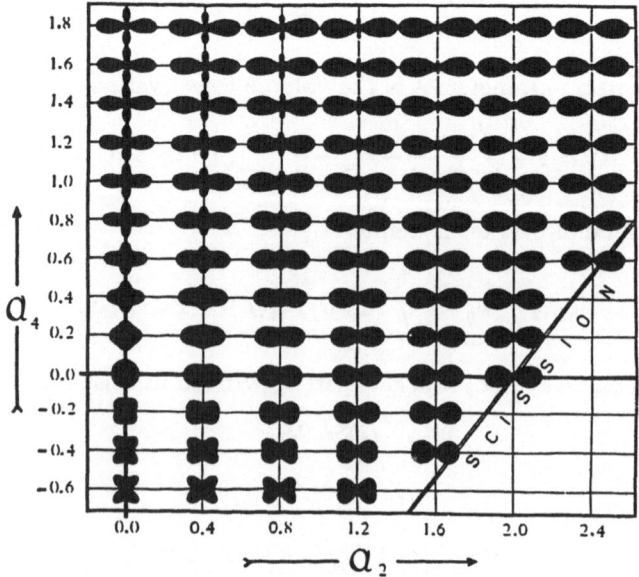

Figure 6.1: A guide to the (α_2, α_4)-distortions (from [Cohen 62]).

Table 6.1: Collection of 3- and 4-symbol brackets (partly taken from [Swiatecki 56b] and [Hasse 78]).

$$(112) = \frac{4}{15} \qquad (123) = \frac{6}{35} \qquad (134) = \frac{8}{63} \qquad (145) = \frac{10}{99}$$

$$(156) = \frac{12}{143} \qquad (167) = \frac{14}{195} \qquad (178) = \frac{16}{255}$$

$$(222) = \frac{4}{35} \qquad (224) = \frac{4}{35} \qquad (233) = \frac{8}{105} \qquad (235) = \frac{20}{231}$$

$$(244) = \frac{40}{693} \qquad (246) = \frac{10}{143} \qquad (255) = \frac{20}{429} \qquad (257) = \frac{42}{715}$$

$$(266) = \frac{28}{715} \qquad (268) = \frac{56}{1105} \qquad (277) = \frac{112}{3315} \qquad (279) = \frac{72}{1615}$$

$$(288) = \frac{48}{1615} \qquad (334) = \frac{4}{77} \qquad (336) = \frac{200}{3003} \qquad (345) = \frac{40}{1001}$$

$$(347) = \frac{70}{1287} \qquad (356) = \frac{14}{429} \qquad (358) = \frac{112}{2431} \qquad (367) = \frac{336}{12155}$$

$$(378) = \frac{504}{20995} \qquad (3,7,10) = \frac{80}{2261} \qquad (444) = \frac{36}{1001} \qquad (446) = \frac{40}{1287}$$

$$(448) = \frac{980}{21879} \qquad (455) = \frac{4}{143} \qquad (457) = \frac{560}{21879} \qquad (466) = \frac{56}{2431}$$

$$(468) = \frac{1008}{46189} \qquad (477) = \frac{4536}{230945} \qquad (488) = \frac{72}{4199}$$

$$(556) = \frac{160}{7293} \qquad (558) = \frac{980}{46189} \qquad (5,5,10) = \frac{1512}{46189} \qquad (567) = \frac{840}{46189}$$

$$(578) = \frac{720}{46189} \qquad (5,7,10) = \frac{1540}{96577} \qquad (5,7,12) = \frac{4752}{185725}$$

$$(666) = \frac{800}{46189} \qquad (668) = \frac{700}{46189} \qquad (677) = \frac{2000}{138567} \qquad (688) = \frac{1200}{96577}$$

$$(778) = \frac{3500}{289731} \qquad (888) = \frac{980}{96577}$$

$$(1111) = \frac{2}{5} \qquad (1113) = \frac{4}{35} \qquad (1122) = \frac{22}{105} \qquad (1124) = \frac{8}{105}$$

$$(1133) = \frac{46}{315} \qquad (1135) = \frac{40}{693} \qquad (1144) = \frac{26}{231} \qquad (1223) = \frac{4}{35}$$

$$(1225) = \frac{4}{77} \qquad (1234) = \frac{284}{3465} \qquad (1236) = \frac{40}{1001} \qquad (1245) = \frac{580}{9009}$$

$$(1333) = \frac{24}{385} \qquad (1335) = \frac{60}{1001} \qquad (1337) = \frac{40}{1287} \qquad (1344) = \frac{136}{3003}$$

$$(2222) = \frac{6}{35} \qquad (2224) = \frac{24}{385} \qquad (2233) = \frac{122}{1155} \qquad (3333) = \frac{482}{5005}$$

moments cf. [Myers 83])

$$\lambda^{-1} = 1 - \frac{1}{5}\alpha_2^2 - \frac{2}{105}\alpha_2^3 + \frac{2}{25}\alpha_2^4 - \frac{2}{35}\alpha_2^2\alpha_4 - \frac{1}{9}\alpha_4^2 - \frac{1}{7}\alpha_3^2$$
$$- \frac{4}{105}\alpha_2\alpha_3^2 + \frac{59}{1225}\alpha_2^2\alpha_3^2 - \frac{20}{231}\alpha_2\alpha_3\alpha_5 - \frac{1}{11}\alpha_5^2 - \frac{2}{77}\alpha_3^2\alpha_4 \cdots$$
$$\alpha_1 = -\frac{27}{35}\alpha_2\alpha_3 - \frac{4}{7}\alpha_3\alpha_4 \cdots \tag{6.17}$$

$$\overline{(\delta r)^2} = \frac{1}{5}\alpha_2^2 - \frac{1}{25}\alpha_2^4 + \frac{1}{9}\alpha_4^2 \cdots$$
$$r_{\text{rms}}^2 = \frac{3}{5}\left(1 + \alpha_2^2 + \frac{10}{21}\alpha_2^3 - \frac{27}{35}\alpha_2^4 + \frac{10}{7}\alpha_2^2\alpha_4 + \frac{5}{9}\alpha_4^2 \cdots\right) \; (\dagger)$$
$$\varrho_n = 1 - \frac{1}{2}\alpha_2 - \frac{1}{5}\alpha_2^2 + \frac{3}{8}\alpha_4 \cdots \; (\dagger) \tag{6.18}$$

$$\zeta_0 = 1 + \alpha_2 - \frac{1}{5}\alpha_2^2 + \alpha_4 \cdots \; (\dagger)$$
$$Q = \frac{8}{5}\pi\left(\alpha_2 + \frac{4}{7}\alpha_2^2 - \frac{1}{7}\alpha_2^3 - \frac{94}{231}\alpha_2^4 + \frac{8}{7}\alpha_2\alpha_4 + \frac{72}{77}\alpha_2^2\alpha_4 + \frac{200}{693}\alpha_4^2 \cdots\right) \; (\dagger)$$
$$Q_4 = \frac{8}{9}\pi\left(\frac{54}{35}\alpha_2^2 + \frac{108}{77}\alpha_2^3 + \frac{8154}{25025}\alpha_2^4 + \alpha_4 + \frac{120}{77}\alpha_2\alpha_4 \cdots\right) \; (\dagger) \tag{6.19}$$

$$\mathcal{J}_\parallel = 1 - \alpha_2 + \frac{3}{7}\alpha_2^2 \cdots \; (*)$$
$$\mathcal{J}_\perp = 1 + \frac{1}{2}\alpha_2 + \frac{9}{7}\alpha_2^2 \cdots \; (*)$$
$$\mathcal{J}_{\text{eff}}^{-1} = \frac{3}{2}\alpha_2 + \frac{6}{7}\alpha_2^2 \cdots \; (*) \tag{6.20}$$

$$B_{\text{Coul}} = 1 - \frac{1}{5}\alpha_2^2 - \frac{4}{105}\alpha_2^3 + \frac{51}{245}\alpha_2^4 - \frac{6}{35}\alpha_2^2\alpha_4 - \frac{5}{27}\alpha_4^2 \cdots$$
$$- \frac{10}{49}\alpha_3^2 - \frac{92}{735}\alpha_2\alpha_3^2 + \frac{2133718}{3112725}\alpha_2^2\alpha_3^2 - \frac{60}{539}\alpha_3^2\alpha_4$$
$$- \frac{5960}{17787}\alpha_2\alpha_3\alpha_5 - \frac{20}{121}\alpha_5^2 \cdots \tag{6.21}$$

$$B_{\text{surf}} = 1 + \frac{2}{5}\alpha_2^2 - \frac{4}{105}\alpha_2^3 - \frac{66}{175}\alpha_2^4 - \frac{4}{35}\alpha_2^2\alpha_4 + \alpha_4^2 \cdots$$
$$+ \frac{5}{7}\alpha_3^2 - \frac{8}{105}\alpha_2\alpha_3^2 - \frac{23736}{13475}\alpha_2^2\alpha_3^2 - \frac{4}{77}\alpha_3^2\alpha_4$$
$$- \frac{40}{231}\alpha_2\alpha_3\alpha_5 + \frac{14}{11}\alpha_5^2 \cdots \tag{6.22}$$

$\dagger$With α_2, α_4 only
$*$With α_2 only

$$B_{\text{curv}} = 1 + \frac{2}{5}\alpha_2^2 + \frac{16}{105}\alpha_2^3 - \frac{82}{175}\alpha_2^4 + \frac{2}{35}\alpha_2^2\alpha_4 + \alpha_4^2 \cdots$$
$$+ \frac{5}{7}\alpha_3^2 + \frac{92}{105}\alpha_2\alpha_3^2 - \frac{41271}{13475}\alpha_2^2\alpha_3^2 + \frac{116}{77}\alpha_3^2\alpha_4$$
$$+ \frac{100}{231}\alpha_2\alpha_3\alpha_5 + \frac{14}{11}\alpha_5^2 \cdots \tag{6.23}$$

$$B_{\text{comp}} = 1 + \frac{4}{5}\alpha_2^2 - \frac{8}{105}\alpha_2^3 - \frac{104}{175}\alpha_2^4 - \frac{8}{35}\alpha_2^2\alpha_4 + 2\alpha_4^2 \cdots$$
$$+ \frac{10}{7}\alpha_3^2 - \frac{16}{105}\alpha_2\alpha_3^2 - \frac{43622}{13475}\alpha_2^2\alpha_3^2 - \frac{8}{77}\alpha_3^2\alpha_4$$
$$- \frac{80}{231}\alpha_2\alpha_3\alpha_5 + \frac{28}{11}\alpha_5^2 \cdots \tag{6.24}$$

$$B_{\text{red}} = 1 + \frac{2}{5}\alpha_2^2 + \frac{16}{105}\alpha_2^3 - \frac{34043}{13475}\alpha_2^4 + \frac{668}{385}\alpha_2^2\alpha_4 + \frac{1330}{891}\alpha_4^2 \cdots$$
$$+ \frac{60}{77}\alpha_3^2 + \frac{1144}{735}\alpha_2\alpha_3^2 - \frac{12432175744}{12371384025}\alpha_2^2\alpha_3^2 + \frac{12200}{5929}\alpha_3^2\alpha_4$$
$$+ \frac{4433580}{847847}\alpha_2\alpha_3\alpha_5 + \frac{26600}{17303}\alpha_5^2 \cdots \tag{6.25}$$

$$B_{\text{sr1}} = 1 - \frac{2}{5}\alpha_2^2 - \frac{4}{105}\alpha_2^3 - \frac{457}{1225}\alpha_2^4 - \frac{8}{105}\alpha_2^2\alpha_4 + \frac{8}{9}\alpha_4^2 \cdots \quad (\dagger)$$
$$B_{\text{sr2}} = 1 + 0\alpha_2^2 + 0\alpha_2^3 - \frac{243}{245}\alpha_2^4 + \frac{4}{7}\alpha_2^2\alpha_4 + \frac{91}{81}\alpha_4^2 \cdots \quad (\dagger)$$
$$B_{\text{v}} = 1 - \frac{1}{5}\alpha_2^2 - \frac{2}{105}\alpha_2^3 - \frac{253}{1225}\alpha_2^4 - \frac{4}{105}\alpha_2^2\alpha_4 + \frac{4}{9}\alpha_4^2 \cdots \quad (\dagger) \tag{6.26}$$

In the natural units of Chap. 1 the relative Coulomb potential Φ and Coulomb potential at the surface Φ_s read $(P_\nu = P_\nu(\cos\theta))$

$$\Phi(r,\theta) = \lambda^{-2}\left[\frac{3}{2} + (\lambda r)^2\left(\frac{3}{5}\alpha_2 P_2 - \frac{1}{2}\right) + \frac{1}{3}(\lambda r)^4\alpha_4 P_4\right.$$
$$\left. + \frac{3}{4}\sum_{k=0}^{4}(k-1)(\lambda r)^k P_k\left\{-\alpha_2^2(22k) - 2\alpha_2\alpha_4(24k) + \frac{1}{3}k\alpha_2^3(222k)\right\}\cdots\right]$$
$$\tag{6.27}$$

$$\Phi_s(\theta) = \lambda^{-2}\left[1 - \frac{2}{5}\alpha_2 P_2 + \alpha_2^2\left(\frac{3}{10} - \frac{3}{35}P_2 + \frac{7}{10}P_2^2 - \frac{9}{35}P_4\right)\right.$$
$$+ \alpha_2^3\left(\frac{3}{35}P_2 - \frac{6}{35}P_2^2 + \frac{3}{5}P_2^3 - \frac{36}{35}P_2 P_4\right) + \alpha_2^4\left(\frac{6}{35}P_2^2 - \frac{3}{35}P_2^3 - \frac{54}{35}P_2 P_4\right)$$
$$+ \alpha_2\alpha_4\left(-\frac{6}{35}P_2 + \frac{23}{15}P_2 P_4\right) + \alpha_2^2\alpha_4\left(-\frac{12}{35}P_2^2 - \frac{6}{35}P_2 P_4\right)$$
$$\left. + \frac{16}{5}P_2^2 P_4 - \frac{36}{35}P_4^2\right) - \frac{2}{3}\alpha_4 P_4 + \alpha_4^2\left(\frac{1}{6} + \frac{5}{6}P_4^2\right)\cdots\right] \tag{6.28}$$

$\dagger$With α_2, α_4 only

51

Fission fragments at the scission point may be polarized due to the Coulomb interaction energy [Hasse 78]. In the following the Coulomb-self energy is also given in the α_n- parameterization for a dipole type inhomogeneous charge density (depending on the parameter ε) of the form

$$\varrho_p(r,\theta) = \overline{\varrho_p}\left[1 - \varepsilon\left(\frac{r}{R_0}P_1(\cos\theta) - \mu_0\right)\right] , \qquad (6.29)$$

where

$$\overline{\varrho_p} = \frac{3Z}{4\pi R_0^3} \qquad (6.30)$$

is the average charge density and

$$\mu_0 = \frac{36}{385}\alpha_3^3 - \frac{72}{175}\alpha_2^2\alpha_3 + \cdots \qquad (6.31)$$

conserves the total charge. The connection between the usual dipole moment

$$D = \int \mathrm{d}^3r\, z\varrho_p(r,\theta) \qquad (6.32)$$

and ε is

$$D = \frac{1}{5}ZeR_0\varepsilon . \qquad (6.33)$$

In writing the Coulomb self energy as

$$E_{\text{Coul,self}} = (1 + \varepsilon\mu_0)^2 E_{\text{Coul,self}}^{\text{hom}} - \varepsilon(1 + \varepsilon\mu_0)E_{\text{Coul,self}}^{\text{inhom,1}} + \varepsilon^2 E_{\text{Coul,self}}^{\text{inhom,2}} , \qquad (6.34)$$

the relative contributions in units of the homogeneous Coulomb energy of the sphere become

$$
\begin{aligned}
B_{\text{Coul,self}}^{\text{hom}} &= \lambda^{-5}\left[1 + \frac{5}{6}\sum p_k q_k + \frac{5}{8}(u_{20} - u_{30} - u_{40})\right] \\
B_{\text{Coul,self}}^{\text{inhom,1}} &= \lambda^{-6}\left[\frac{5}{6}\sum(q_k s_k + p_k t_k) + \frac{5}{4}\left(\frac{6}{5}u_{11} + u_{21} - \frac{2}{3}u_{31} - \frac{3}{2}u_{41}\right)\right] \\
B_{\text{Coul,self}}^{\text{inhom,2}} &= \lambda^{-7}\left[\frac{1}{21} + \frac{5}{6}\sum s_k t_k + \frac{1}{4}\left(u_{12} + \frac{1}{2}u_{22} - \frac{5}{2}u_{32} - 5u_{42}\right)\right] .
\end{aligned}
\qquad (6.35)
$$

Here, we defined the reduced multipole moments

$$
\begin{aligned}
p_k(\alpha) &= \frac{3}{2}\int \mathrm{d}x\, P_k(x)\frac{[1 + \alpha(x)]^{2-k}}{2 - k} \\
q_k(\alpha) &= \frac{3}{2}\int \mathrm{d}x\, P_k(x)\frac{[1 + \alpha(x)]^{3+k}}{3 + k} \\
s_k(\alpha) &= \frac{3}{2}\int \mathrm{d}x\, P_1(x)\,P_k(x)\frac{[1 + \alpha(x)]^{3-k}}{3 - k} \\
t_k(\alpha) &= \frac{3}{2}\int \mathrm{d}x\, P_1(x)\,P_k(x)\frac{[1 + \alpha(x)]^{4+k}}{4 + k} \\
u_{kl}(\alpha) &= \int \mathrm{d}x\, [P_1(x)]^l\,[\alpha(x)]^k ,
\end{aligned}
\qquad (6.36)
$$

where

$$\alpha(\cos\theta) = \sum \alpha_n P_n(\cos\theta) \tag{6.37}$$

is the deviation of the shape from the sphere. The quantities p_2 and s_3 are defined by their series expansions and they have the special form

$$p_2(\alpha) = \frac{3}{2}\int dx\, P_2 \log[1+\alpha(x)]$$

$$s_3(\alpha) = \frac{3}{2}\int dx\, P_3 P_1 \log[1+\alpha(x)], \tag{6.38}$$

respectively. All integrals run from -1 up to $+1$. A list of the reduced multipoles p_k, q_k, s_k, t_k is given in Table 6.2 and one gets

$$\lambda^{-1} = q_0^{-1/3}$$

$$\mu_0 = \lambda^{-1}q_1/q_0 \tag{6.39}$$

and the Coulomb potential (6.27) becomes

$$\Phi(r,\theta) = \lambda^{-2}\left[\frac{3}{2} - \frac{1}{2}(\lambda r)^2 + \sum_{k=0}^{\infty}(\lambda r)^k p_k P_k\right]. \tag{6.40}$$

The reduced multipoles u_{kl} can be evaluated in terms of the n-symbol brackets (6.8). Up to quadratic order in α_2, α_3, α_4 and in the dipole distortion parameter ε, (α_1 is eliminated by center-of-mass conservation) the relative Coulomb self energy reads

$$\begin{aligned}
B_{\text{Coul,self}} = {}& 1 - \frac{1}{5}\alpha_2^2 - \frac{4}{105}\alpha_2^3 - \frac{10}{49}\alpha_3^2 - \frac{92}{735}\alpha_2\alpha_3^2 + \frac{51}{245}\alpha_2^4 + \frac{2133718}{3112725}\alpha_2^2\alpha_3^2 \\
& + \varepsilon\alpha_3\left(\frac{12}{49}\alpha_2 + \frac{40}{847}\alpha_3^2 + \frac{8}{49}\alpha_2^2\right) \\
& + \varepsilon^2\left(\frac{1}{21} + \frac{2}{15}\alpha_2 + \frac{128}{735}\alpha_2^2 + \frac{491}{6615}\alpha_3^2\right)\cdots
\end{aligned} \tag{6.41}$$

The hydrodynamic mass parameters and viscosity [Chandrasekhar 59] and one body dissipation coefficients discussed in Sect. 1.8 read

$$B_{nn} = \frac{3}{n(2n+1)}$$

$$Z_{nn} = 2\frac{n-1}{n}$$

$$D_{nn} = \frac{1}{2n+1} \tag{6.42}$$

and in the pure LDM the stiffness and squared eigenfrequency of Sect. 1.8 around the spherical shape for all n become

$$C_{nn} = \frac{n-1}{2n+1}\left[(n+2) - \frac{20y}{2n+1}\right]$$

$$\omega_n^2 = \frac{n(n-1)}{3}\left[(n+2) - \frac{20y}{2n+1}\right]. \tag{6.43}$$

Here $y = 1 - x$ and x is the fissility of Eq. (1.75).

Table 6.2: Some values of the reduced multipoles used to calculate Coulomb self and interaction energies (from [Hasse 78]).

$$p_0 = \frac{3}{10}\alpha_2^2 + \frac{13}{14}\alpha_3^2 + \frac{1}{6}\alpha_4^2 + \frac{3}{22}\alpha_5^2 + \frac{729}{1225}\alpha_2^2\alpha_3^2$$

$$p_1 = -\frac{27}{35}\alpha_2\alpha_3$$

$$p_2 = \frac{3}{5}\alpha_2 - \frac{3}{35}\alpha_2^2 - \frac{2}{35}\alpha_3^2$$

$$p_3 = \frac{3}{7}\alpha_3 - \frac{8}{35}\alpha_2\alpha_3$$

$$p_4 = -\frac{9}{35}\alpha_2^2 - \frac{9}{77}\alpha_3^2 + \frac{1}{3}\alpha_4$$

$$q_0 = 1 + \frac{3}{5}\alpha_2^2 + \frac{3}{7}\alpha_3^2 + \frac{2}{35}\alpha_2^3 + \frac{4}{35}\alpha_2\alpha_3^2 + \frac{243}{1225}\alpha_2^2\alpha_3^2$$
$$+ \frac{1}{3}\alpha_4^2 + \frac{6}{35}\alpha_2^2\alpha_4 + \frac{3}{11}\alpha_5^2 + \frac{6}{77}\alpha_3^2\alpha_4 + \frac{20}{77}\alpha_2\alpha_3\alpha_5$$

$$q_1 = -\frac{72}{175}\alpha_2^2\alpha_3 + \frac{36}{385}\alpha_3^3 + \frac{2118}{13475}\alpha_2\alpha_3^2 + \frac{4}{7}\alpha_3\alpha_4$$

$$q_2 = \frac{3}{5}\alpha_2 + \frac{12}{35}\alpha_2^2 - \frac{8}{35}\alpha_3^2$$

$$q_3 = \frac{3}{7}\alpha_3 + \frac{4}{7}\alpha_2\alpha_3$$

$$q_4 = \frac{18}{35}\alpha_2^2 + \frac{18}{77}\alpha_3^2 + \frac{1}{3}\alpha_4$$

$$q_5 = \frac{10}{11}\alpha_2\alpha_3 + \frac{3}{11}\alpha_5$$

$$q_6 = \frac{400}{1001}\alpha_3^2$$

$$s_0 = -\frac{9}{35}\alpha_2\alpha_3 - \frac{78}{175}\alpha_2^2\alpha_3 + \frac{12}{385}\alpha_3^3 + \frac{8}{21}\alpha_3\alpha_4$$

$$s_1 = \frac{2}{5}\alpha_2 + \frac{11}{70}\alpha_2^2 + \frac{23}{210}\alpha_3^2$$

$$s_2 = \frac{9}{35}\alpha_3 - \frac{54}{175}\alpha_2\alpha_3$$

$$s_3 = \frac{9}{35}\alpha_2 - \frac{3}{35}\alpha_2^2 - \frac{18}{385}\alpha_3^2 + \frac{4}{21}\alpha_4$$

$$s_4 = \frac{4}{21}\alpha_3$$

Table 6.2: (cont.)

$$t_0 = -\frac{72}{175}\alpha_2^2\alpha_3 + \frac{36}{385}\alpha_3^3 + \frac{2118}{13475}\alpha_2\alpha_3^2 + \frac{4}{7}\alpha_3\alpha_4$$

$$t_1 = \frac{1}{5} + \frac{2}{5}\alpha_2 + \frac{22}{35}\alpha_2^2 + \frac{46}{105}\alpha_3^2$$

$$t_2 = \frac{9}{35}\alpha_3 + \frac{96}{175}\alpha_2\alpha_3$$

$$t_3 = \frac{9}{35}\alpha_2 + \frac{18}{35}\alpha_2^2 + \frac{108}{385}\alpha_3^2 + \frac{4}{21}\alpha_4$$

$$t_4 = \frac{4}{21}\alpha_3 + \frac{142}{165}\alpha_2\alpha_3$$

$$t_5 = \frac{24}{77}\alpha_2^2 + \frac{360}{1001}\alpha_3^2 + \frac{5}{33}\alpha_4$$

$$t_6 = \frac{540}{1001}\alpha_2\alpha_3 + \frac{18}{143}\alpha_5$$

$$t_7 = \frac{100}{429}\alpha_3^2$$

6.2.2 a_n-Parameterization

The volume conservation can be incorporated into the deformation parameters [Swiatecki 56a]

$$R(\theta) = R_0 \left(1 + \sum_{n=0}^{N} a_n P_n(\cos\theta)\right) \tag{6.44}$$

so that

$$1 + a_0 = \lambda^{-1}$$
$$a_n = \lambda^{-1}\alpha_n \quad \text{for } n \geq 1 . \tag{6.45}$$

With this conversion, all formulae of Sect. 6.2.1 remain valid.

In fact, most of the results of Sect. 6.2.1 have been derived originally in the a_n-parameterization [Nossoff 55], [Swiatecki 56a], [Swiatecki 58] and then transformed into the α_n-parameterization by eliminating a_0. The relative surface energy can be written as

$$B_{\text{surf}} = 1 + \frac{1}{2}\sum_{n=2}^{\infty}\frac{(n-1)(n+2)}{2n+1}a_n^2 + \sum_{mn=2}^{\infty}\left[\frac{1}{(2n+1)(2m+1)}a_n^2 a_m^2\right.$$

$$+ \frac{36(m+1)(n+1)}{(2m+1)(2m+3)(2n+1)(2n+3)}a_m a_{m+1} a_n a_{n+1}\right]$$

$$-\frac{1}{3}\sum_{lmn=2}^{\infty}(lmn)a_l a_m a_n$$

$$-\frac{1}{128}\sum_{klmn=2}^{\infty}\sum_{i=0}^{\infty}\left[k(k+1) + l(l+1) - i(i+1)\right]\left[m(m+1)\right.$$

$$+ n(n+1) - i(i+1)\right](2i+1)(ikl)(imn)a_k a_l a_m a_n . \tag{6.46}$$

55

The fourth order term of the relative Coulomb energy has been given incorrectly in [Nossoff 55] and was later corrected in [Swiatecki 58],

$$
\begin{aligned}
B_{\text{Coul}} \; = \; & 1 - 5 \sum_{n=2}^{\infty} \frac{(n-1)}{(2n+1)^2} a_n^2 + 5 \sum_{mn=2}^{\infty} \left[\frac{n-4}{(2n+1)^2(2m+1)} a_n^2 a_m^2 \right. \\
& + \frac{90(4n^2 - n - 6)(m+1)(n+1)}{(2m+1)(2m+3)(2n+1)^2(2n+3)^2} a_m a_{m+1} a_n a_{n+1} \Bigg] \\
& - \frac{5}{12} \sum_{lmn=2}^{\infty} \frac{7n-10}{2n+1} (lmn) a_l a_m a_n \\
& + \frac{15}{16} \sum_{klmn=2}^{\infty} \sum_{i=0}^{\infty} \left\{ \left[i(i+1) - m(m+1) + \frac{1}{3}(n^2 - n + 1) \right] \right. \\
& \times \frac{2i+1}{2n+1} + \left[\frac{4}{9} m(m+1) - \frac{1}{2} i(i+1) + \frac{2}{9} n(n+1) \right] \\
& \times (2i+1) - \frac{1}{2}(i-1)(i+2) \bigg\} (ikl)(imn) a_k a_l a_m a_n \, .
\end{aligned}
\tag{6.47}
$$

Furthermore,

$$
\overline{(\delta r)^2} = \sum_{n=0}^{\infty} \frac{1}{2n+1} a_n^2 \, .
\tag{6.48}
$$

The hydrodynamic mass parameter of Sect. 1.8 with respect to a_2 reads [Foland 59]

$$
B_{a_2} = \frac{3}{10} \left(1 + \frac{9}{7} a_2 \cdots \right)
\tag{6.49}
$$

6.2.3 α_{n0}-Parameterization

Instead of Legendre polynomials axially symmetric spherical harmonics have been employed [Wilets 64],

$$
R(\theta) = \lambda^{-1} R_0 \left(1 + \sum_{n=1}^{N} \alpha_{n0} Y_{n0}(\theta) \right) \, .
\tag{6.50}
$$

With the conversion

$$
\alpha_{n0} = \sqrt{\frac{4\pi}{2n+1}} \, \alpha_n \, ,
\tag{6.51}
$$

all formulae of Sect. 6.2.1 remain valid.

6.2.4 β_n-Parameterization

This parameterization is obtained if volume conservation is taken into account only up to quadratic terms in the deformation parameters [Bohr 75],

$$
R(\theta) = R_0 \left[1 + \sum_{n=2}^{N} \left(\beta_n Y_{n0}(\theta) - \frac{\beta_n^2}{4\pi} \right) \right] \, .
\tag{6.52}
$$

Up to quadratic order $\beta_n = \alpha_{n0}$. The parameter β_2 usually is denoted by β.

In the natural units of Chap. 1 the relative quadrupole and hexadecapole moments are [Brack 74]

$$Q = 4\sqrt{\frac{\pi}{5}}\left(\beta_2 + 0.360\beta_2^2 + 0.967\beta_2\beta_4 + 0.328\beta_4^2\right.$$
$$\left. + 0.023\beta_2^3 - 0.021\beta_2^4 + 0.499\beta_2^2\beta_4 \cdots\right) \tag{6.53}$$

$$Q_4 = \frac{4}{3}\sqrt{\pi}\left(\beta_4 + 0.725\beta_2^2 + 0.983\beta_2\beta_4 + 0.411\beta_4^2\right.$$
$$\left. + 0.416\beta_2^3 + 1.656\beta_2^4\beta_4 + 0.055\beta_2^4 \cdots\right). \tag{6.54}$$

The Krappe-Nix finite range surface energy of Sect. 1.4.2 has been evaluated in [Krappe 73] in the β_n-parameterization,

$$B_{KN} = 1 + \frac{1}{4\pi}\sum_{n=2}^{\infty}\beta_n^2 C_n^{(KN)} \tag{6.55}$$

with

$$C_n^{(KN)} = (\sigma + 1)\left[\sigma - 1 + (\sigma + 1)e^{-2\sigma}\right]$$
$$- 2\sigma^3 I_{n+1/2}(\sigma)\,K_{n+1/2}(\sigma), \tag{6.56}$$

where $\sigma = R_0/a$ and I_n, K_n are modified Bessel functions. Up to the orders of σ^{-4} and $\exp(-2\sigma)$ the expansion becomes

$$C_n^{(KN)} = \frac{1}{2}n(n+1) - 1 - \frac{3}{8}(n-1)n(n+1)(n+2)\sigma^{-2}. \tag{6.57}$$

Similarly, for the Yukawa-plus-exponential folding energy of Sect. 1.4.2

$$B_{YE} = 1 + \frac{1}{4\pi}\sum_{n=2}^{\infty}\beta_n^2 C_n^{(YE)}$$
$$C_n^{(YE)} = (\sigma + 1)(2\sigma^2 + \sigma + 1)e^{-2\sigma} - (\sigma^2 + 1)$$
$$+ 2\frac{\partial}{\partial(1/\sigma)}\left(\sigma^3 j_n(i\sigma)\,h_n^{(1)}(i\sigma)\right), \tag{6.58}$$

where $j_n(i\sigma), h_n^{(1)}(i\sigma)$ are spherical Bessel and Hankel functions of imaginary argument, respectively. For $n = 2$, one gets

$$C_2^{(YE)} = 2 - \frac{27}{\sigma^2} + \left[4\sigma^3 + 14\sigma^2 + 32\sigma + 52 + \frac{54}{\sigma} + \frac{27}{\sigma^2}\right]e^{-2\sigma}. \tag{6.59}$$

6.2.5 β_q-Parameterization

Similarly, the β_q-parameterization, cf. [Löbner 70], can be used for quadrupole deformation if volume conservation is taken into account for the quadrupoloid,

$$R(\theta) = R_0[1 + \alpha_{00}Y_{00} + \beta_q Y_{20}(\theta)] \tag{6.60}$$

$$\alpha_{00}Y_{00} = -\left(\frac{\beta_q}{2\sqrt{\pi}}\right)^2 - \frac{2\sqrt{5}}{21}\left[\left(\frac{\beta_q}{2\sqrt{\pi}}\right)^3 + \left(\frac{\beta_q}{2\sqrt{\pi}}\right)^5\right] + \frac{127}{441}\left(\frac{\beta_q}{2\sqrt{\pi}}\right)^6 \cdots \tag{6.61}$$

57

where $Y_{00} = (4\pi)^{-1/2}$. In this respect, it is a combination of the α_n -, a_n - and α_{n0}- parameterizations.

6.3 Distorted Spheroids

6.3.1 ε_n-Parameterization

Higher than spheroidal deformations can also be obtained by expansions about a spheroid. Nilsson [Nilsson 69], cf. also the recent paper [Böning 87], defined a deformed oscillator potential as a generalization of Eq. (5.17),

$$V_{\text{osc}} = \frac{\hbar}{2}\omega_0(\varepsilon_1, \cdots, \varepsilon_N)r_t^2 \left[1 - \frac{2}{3}\varepsilon_2 P_2(\cos\theta_t) + 2\sum_{n=1,\neq 2}^{N} \varepsilon_n P_n(\cos\theta_t) \right] , \quad (6.62)$$

where the stretched coordinates (r_t, θ_t) are defined in the same way as in Chap. 5 but with $\omega_0(\varepsilon)$ replaced by $\omega_0(\varepsilon_1, \cdots, \varepsilon_N)$. Then the volume conservation can no longer be obtained in closed form,

$$\frac{\omega_0^{3/2}(\varepsilon_1, \cdots, \varepsilon_N)\omega_z^{1/2}\omega_\perp}{\overset{o}{\omega}_0^3}$$

$$= \frac{1}{2}\int_{-1}^{1} d\mu \left[1 - \frac{2}{3}\varepsilon_2 P_2(\mu) + 2\sum_{n=1,\neq 2}^{N} \varepsilon_n P_n(\mu) \right]^{-3/2} . \quad (6.63)$$

Here ω_z and $\omega_\perp$ are also defined as in Eq. (5.20) but with $\omega_o(\varepsilon)$ replaced by $\omega_0(\varepsilon_1, \cdots, \varepsilon_N)$. Also, if odd ε_i are involved, ε_1 must be eliminated numerically by center of mass conservation.

The corresponding equipotential surfaces $R(\theta)$ are obtained from

$$R^2(\theta) = R_0^2 \frac{\overset{o}{\omega}_0^2 \omega_0}{\omega_z\omega_\perp^2} \frac{1 - \frac{1}{3}\varepsilon_2 + \frac{2}{3}\varepsilon_2 P_2(\cos\theta_t)}{1 - \frac{2}{3}\varepsilon_2 P_2(\cos\theta_t) + 2\sum\limits_{n=1,\neq 2}^{N} \varepsilon_n P_n(\cos\theta_t)} \quad (6.64)$$

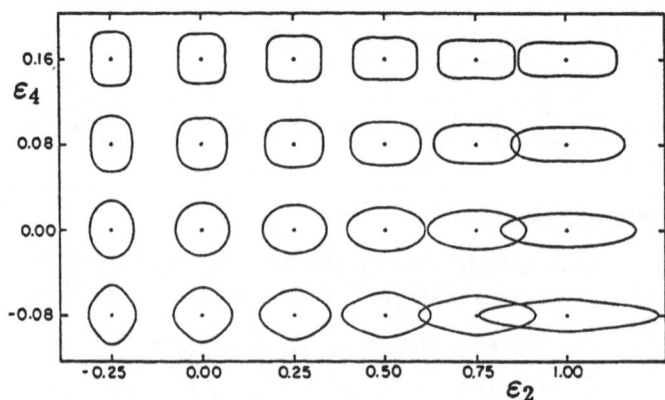

Figure 6.2: Shapes in the $(\varepsilon_2 - \varepsilon_4)$-parameterization (from [Nilsson 69]).

58

and using (5.30) to replace θ_t in terms of θ. Shapes contained in the ε_n-parameterization are displayed in Fig. 6.2. The parameter ε_2 is also sometimes denoted by ε. Other methods of distorting a spheroidal potential by Gaussians are available [Chasman 70], [Jänecke 72b] but did not gain widespread attention.

6.3.2 Spheroidal Coordinates

Swiatecki and others [Businaro 55a], [Swiatecki 56a], [Swiatecki 58], cf. also [Moon 61] used prolate spheroidal coordinates (ξ, η) defined in terms of cylindrical coordinates by

$$\varrho = k\sqrt{(1 - \xi^2)(1 - \eta^2)}$$
$$z = k\xi\eta, \tag{6.65}$$

where
$$k = ec \tag{6.66}$$

and, e, c, are the eccentricity and major semiaxis of Chap. 5, respectively. The spheroid is then given by $\eta(\xi) = e^{-1}$ and harmonically distorted spheroids read

$$\eta(\xi) = e^{-1}\left(1 + \sum_{n=0}^{N} \alpha_n P_n(\xi)\right). \tag{6.67}$$

Volume conservation is achieved with Eq. (5.3) and by setting α_0 to the value satisfying the equation

$$(3 - e^2)\alpha_0 - \frac{2}{5}e^2\alpha_2 + 3\sum_{n=0}^{N}\frac{\alpha_n^2}{2n+1} + \frac{1}{2}\sum_{l,m,n=0}^{N}(lmn)\alpha_l\alpha_m\alpha_n = 0. \tag{6.68}$$

Here, (lmn) are the three-symbol brackets of Sect. 6.2.1. In case of odd distortions, constancy of the center-of-mass is achieved by setting α_1 to the value satisfying the equation

$$c_{11} + \frac{3}{2}c_{12} + c_{13} + \frac{1}{4}c_{14} - \frac{3}{5}e^2\left(c_{11} + \frac{2}{7}c_{31} + \frac{1}{2}c_{12} + \frac{1}{7}c_{32}\right) = 0 \tag{6.69}$$

The coefficients c_{ij} are obtained from the recursion relation

$$c_{i,j+1} = \frac{2i+1}{2}\sum_{l,m=0}^{N}(ilm)\alpha_m c_{lj}$$

$$c_{i0} = \delta_{i,0}, \quad c_{i1} = \alpha_i. \tag{6.70}$$

Up to second order in the parameters, Eq. (6.69) reads

$$\alpha_1(5 - 3e^2) - \frac{6}{7}e^2\alpha_3 + \frac{3}{2}\sum_{l,m=0}^{N}\alpha_{lm}\left[\frac{3}{2}(5 - e^2)(1lm) - (3lm)\right]\cdots = 0 \tag{6.71}$$

$$\alpha_1 = \frac{6}{7}\frac{e^2}{5 - 3e^2}\alpha_3\cdots$$

Expressions for Coulomb and surface energies can be found in [Swiatecki 58] and the total deformation energy is [Businaro 55a] ($\alpha = e^2/3$ and x is the fissility)

$$
\begin{aligned}
B_{\text{Def}} = {}& \frac{2}{5}(1-x)\alpha^2 + \frac{4}{105}(29-32x)\alpha^3 + \frac{1}{35}(101-116x)\alpha^4 \\
& + \frac{2}{35}(1-8x)\alpha^2\alpha_4 + \frac{1}{27}(27-10x)\alpha_4^2 + \frac{5}{49}(7-4x)\alpha_3^2 \\
& + \frac{8}{245}(56-41x)\alpha\alpha_3^2 + \frac{4}{5}(1-x)\alpha\alpha_2 + \frac{12}{35}(6-7x)\alpha^2\alpha_2 \\
& + \frac{8}{1225}(889-1150x)\alpha^3\alpha_2 + \frac{4}{35}(8-11x)\alpha\alpha_2^2 \\
& + \frac{2}{1225}(1897-2995x)\alpha^2\alpha_2^2 + \frac{6}{29645}(25641-20936x)\alpha^2\alpha_3^2 \\
& + \frac{48}{35}(1-x)\alpha\alpha_1\alpha_3 - \frac{12}{245}(7+8x)\alpha_1\alpha_2\alpha_3 \cdots
\end{aligned}
\tag{6.72}
$$

which for small values of $(1-x)$ simply becomes

$$
B_{\text{Def}} = \frac{2}{5}(\alpha+\alpha_2)^2 \left[(1-x) + \frac{2}{7}(\alpha+\alpha_2)\cdots\right].
\tag{6.73}
$$

The transformation from the coordinates α, α_n of Businaro to the coordinates a_n of Sect. 6.2.2 is

$$
\begin{aligned}
a_1 &= \alpha_1 + \alpha\alpha_1 - \frac{9}{7}\alpha\alpha_3 \cdots \\
a_2 &= \alpha + \alpha_2 + \frac{10}{7}\alpha^2 + \frac{9}{7}\alpha\alpha_2 - \frac{12}{7}\alpha\alpha_4 \cdots \\
a_3 &= \alpha_3 + \frac{4}{3}\alpha\alpha_3 + \frac{20}{11}\alpha\alpha_5 \cdots \\
a_4 &= \alpha_4 + \frac{27}{35}\alpha^2 + \frac{18}{35}\alpha\alpha_2 + \frac{86}{77}\alpha\alpha_4 - \frac{36}{143}\alpha\alpha_6 \cdots \\
a_5 &= \alpha_5 + \frac{20}{21}\alpha\alpha_3 + \frac{119}{117}\alpha\alpha_5 + \frac{213}{65}\alpha\alpha_7 \cdots
\end{aligned}
\tag{6.74}
$$

6.4 Relations Between Small Shape Parameters

By comparing the ratios of axes in the ε parameterization, Eq. (5.23), in the α_2 parameterization, Eq. (5.34), and in the δ parameterization, Eq. (5.34), one derives the conversions [Arseniev 68]

$$
\begin{aligned}
\varepsilon &= \frac{3\alpha_2}{2+\alpha_2} \cdots \\
\alpha_2 &= \frac{2\varepsilon}{3-\varepsilon} \cdots \\
\delta &= \varepsilon \frac{1-\frac{1}{6}\varepsilon}{1+\frac{2}{9}\varepsilon^2}.
\end{aligned}
\tag{6.75}
$$

Series expansions of the small shape parameters β_q (Sect. 6.2.5), $\beta = \beta_2$ (Sect. 6.2.4), $\varepsilon = \varepsilon_2$ (Sect. 5.2), δ (Eq. (5.33)) in terms of each other and $(\eta\kappa)$ (Eq. (5.28)) are given in Table 6.3 (from [Löbner 70]). Other interrelations can be derived by expanding the appropriate volume conservation factors. The ε_n parameterization (Sect. 6.3.1) is often employed also for larger values of the parameters. Then series expansions are inappropriate and the conversion from (β_2, β_4) to $(\varepsilon_2, \varepsilon_4)$ is shown in Fig. 6.3. The transformation from $(\varepsilon_2, \varepsilon_4)$ to the a_n-parameterization is [Seeger 75]

$$
\begin{aligned}
a_2 &= \frac{2}{3}\varepsilon_2 + \frac{5}{63}\varepsilon_2^2 + \frac{2}{21}\varepsilon_2\varepsilon_4 + \frac{50}{231}\varepsilon_4^2 \cdots \\
a_4 &= -\varepsilon_4 + \frac{12}{35}\varepsilon_2^2 + \frac{50}{231}\varepsilon_2\varepsilon_4 + \frac{243}{1001}\varepsilon_4^2 \cdots \\
a_6 &= -\frac{40}{33}\varepsilon_2\varepsilon_4 + \frac{10}{33}\varepsilon_4^2 \cdots \\
a_8 &= \frac{245}{429}\varepsilon_4^2 \cdots
\end{aligned}
\tag{6.76}
$$

which is shown in Fig. 6.4.

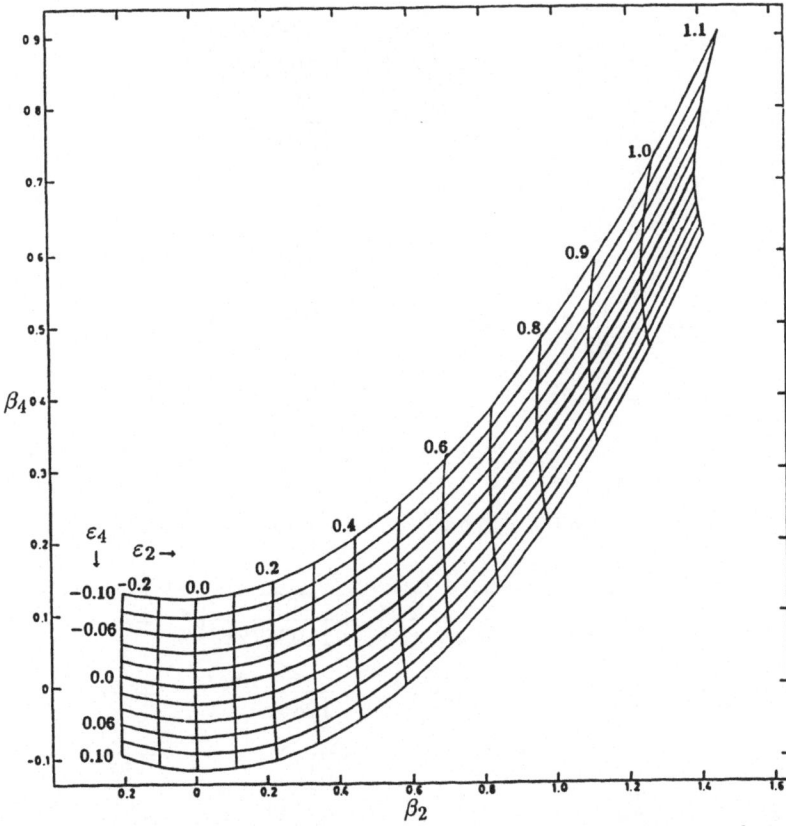

Figure 6.3: Relations between the coordinates $(\varepsilon_2, \varepsilon_4)$ and (β_2, β_4) (from [Nilsson 69] Note that in the original figure the sign of ε_4 is reversed).

Table 6.3: Relations between different deformation parameters (from [Löbner 70]).

$$\beta_q = \beta - \frac{9}{56}\sqrt{\frac{5}{\pi}}\beta^2 + \frac{397}{392\pi}\beta^3 - \frac{367751}{724416\pi}\sqrt{\frac{5}{\pi}}\beta^4 + \cdots$$

$$= \sqrt{\frac{\pi}{5}}\left(\frac{4}{3}\varepsilon + \frac{10}{63}\varepsilon^2 + \frac{2896}{6615}\varepsilon^3 + \frac{519754}{4584195}\varepsilon^4 + \cdots\right)$$

$$= \sqrt{\frac{\pi}{5}}\left(\frac{4}{3}\delta + \frac{2}{21}\delta^2 + \frac{5696}{6615}\delta^3 + \frac{2994688}{4584195}\delta^4 + \cdots\right)$$

$$= \sqrt{\frac{\pi}{5}}\left(\frac{4}{3}(\eta\kappa) + \frac{10}{63}(\eta\kappa)^2 + \frac{1916}{6615}(\eta\kappa)^3 + \frac{207134}{4584195}(\eta\kappa)^4 + \cdots\right)$$

$$\beta = \beta_q + \frac{9}{56}\sqrt{\frac{5}{\pi}}\beta_q^2 - \frac{169}{224\pi}\beta_q^3 - \frac{167561}{827904\pi}\sqrt{\frac{5}{\pi}}\beta_q^4 + \cdots$$

$$= \sqrt{\frac{\pi}{5}}\left(\frac{4}{3}\varepsilon + \frac{4}{9}\varepsilon^2 + \frac{4}{27}\varepsilon^3 + \frac{4}{81}\varepsilon^4 + \cdots\right)$$

$$= \sqrt{\frac{\pi}{5}}\left(\frac{4}{3}\delta + \frac{2}{3}\delta^2 + \frac{2}{3}\delta^3 + \frac{11}{18}\delta^4 + \cdots\right)$$

$$= \sqrt{\frac{\pi}{5}}\left(\frac{4}{3}(\eta\kappa) + \frac{4}{9}(\eta\kappa)^2 - \frac{20}{243}(\eta\kappa)^4 + \cdots\right)$$

$$\varepsilon = \frac{3}{4}\sqrt{\frac{5}{\pi}}\beta_q - \frac{75}{224\pi}\beta_q^2 - \frac{81}{128\pi}\sqrt{\frac{5}{\pi}}\beta_q^3 + \frac{890405}{1103872\pi^2}\beta_q^4 + \cdots$$

$$= \frac{3}{4}\sqrt{\frac{5}{\pi}}\beta - \frac{15}{16\pi}\beta^2 + \frac{15}{64\pi}\sqrt{\frac{5}{\pi}}\beta^3 - \frac{75}{256\pi^2}\beta^4 + \cdots$$

$$= \delta + \frac{1}{6}\delta^2 + \frac{5}{18}\delta^3 + \frac{37}{216}\delta^4 + \cdots$$

$$= (\eta\kappa) - \frac{1}{9}(\eta\kappa)^3 - \frac{2}{81}(\eta\kappa)^4 + \frac{1}{81}(\eta\kappa)^5$$

$$\delta = \frac{3}{4}\sqrt{\frac{5}{\pi}}\beta_q - \frac{45}{56\pi}\beta_q^2 - \frac{57}{56\pi}\sqrt{\frac{5}{\pi}}\beta_q^3 + \frac{43115}{17248\pi^2}\beta_q^4 + \cdots$$

$$= \frac{3}{4}\sqrt{\frac{5}{\pi}}\beta - \frac{45}{32\pi}\beta^2 + \frac{675}{512\pi^2}\beta^4 + \cdots$$

$$= \varepsilon - \frac{1}{6}\varepsilon^2 - \frac{2}{9}\varepsilon^3 + \frac{1}{27}\varepsilon^4 + \cdots$$

$$= (\eta\kappa) - \frac{1}{6}(\eta\kappa)^2 - \frac{1}{3}(\eta\kappa)^3 + \frac{4}{81}(\eta\kappa)^4 + \frac{35}{243}(\eta\kappa)^5$$

$$\eta\kappa = \frac{3}{4}\sqrt{\frac{5}{\pi}}\beta_q - \frac{75}{224\pi}\beta_q^2 - \frac{51}{128\pi}\sqrt{\frac{5}{\pi}}\beta_q^3 + \frac{759505}{1103872\pi^2}\beta_q^4 + \cdots$$

$$= \frac{3}{4}\sqrt{\frac{5}{\pi}}\beta - \frac{15}{16\pi}\beta^2 + \frac{15}{32\pi}\sqrt{\frac{5}{\pi}}\beta^3 - \frac{125}{128\pi^2}\beta^4 + \cdots$$

$$= \varepsilon + \frac{1}{9}\varepsilon^3 + \frac{2}{81}\varepsilon^4 + \frac{2}{81}\varepsilon^5 + \cdots$$

$$= \delta + \frac{1}{6}\delta^2 + \frac{7}{18}\delta^3 + \frac{163}{648}\delta^4 + \cdots$$

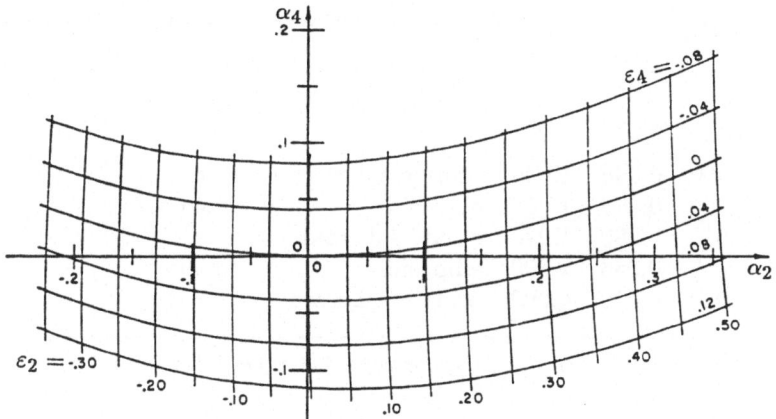

Figure 6.4: Relation between the coordinates (α_2, α_4) and $(\varepsilon_2, \varepsilon_4)$ (from [Seeger 75]).

6.5 Triaxial Shapes

6.5.1 Triaxial Ellipsoids

General definition:

$$\frac{x^2}{a^2} + \frac{y^2}{b^2} + \frac{z^2}{c^2} = 1 , \tag{6.77}$$

to be solved, for instance, for $x_s(y, z)$.
Volume conservation:

$$a\,b\,c = R_0^3 . \tag{6.78}$$

To take volume conservation into account explicitly, Hill and Wheeler [Hill 53] introduced the parameters α_H, β_H,

$$
\begin{aligned}
a &= R_0 \exp\left[\alpha_H \cos\left(\gamma_H - \frac{2\pi}{3}\right)\right] \\
b &= R_0 \exp\left[\alpha_H \cos\left(\gamma_H + \frac{2\pi}{3}\right)\right] \\
c &= R_0 \exp\left[\alpha_H \cos\gamma_H\right] ,
\end{aligned} \tag{6.79}
$$

where α_H is unrestricted and $0 \le \gamma_H < \pi/3$.
Other notation:

$$\beta_H = \sqrt{4\pi/5}\, \alpha_H . \tag{6.80}$$

Special cases of γ_H are listed in Table 6.4 and the symmetries in the (β_H, γ_H) plane are displayed in Fig. 6.5. Some quantities have been derived in [Myers 66] up to the order α_H^2,

$$
\begin{aligned}
\overline{(\delta r)^2} &= \frac{1}{5}\alpha_H^2 \left(1 - \frac{1}{7}\alpha_H \cos 3\gamma_H \cdots\right) \\
B_{\text{surf}} &= 1 + \frac{2}{5}\alpha_H^2 - \frac{2}{21}\alpha_H^3 \cos 3\gamma_H \cdots \\
B_{\text{Coul}} &= 1 - \frac{1}{5}\alpha_H^2 - \frac{1}{105}\alpha_H^3 \cos 3\gamma_H \cdots
\end{aligned} \tag{6.81}
$$

63

					Symmetry
γ_H	a/R_0	b/R_0	c/R_0	Shape	axis
0°	0.990	0.990	1.020	prolate spheroid	z
30°	1.000	0.983	1.017	ellipsoid	none
60°	1.010	0.980	1.010	oblate spheroid	y
90°	1.017	0.983	1.000	ellipsoid	none
120°	1.020	0.990	0.990	prolate spheroid	x

With further increase of γ the cycle repeats, except for cyclic interchange of the labels x, y, z, a, b, c.

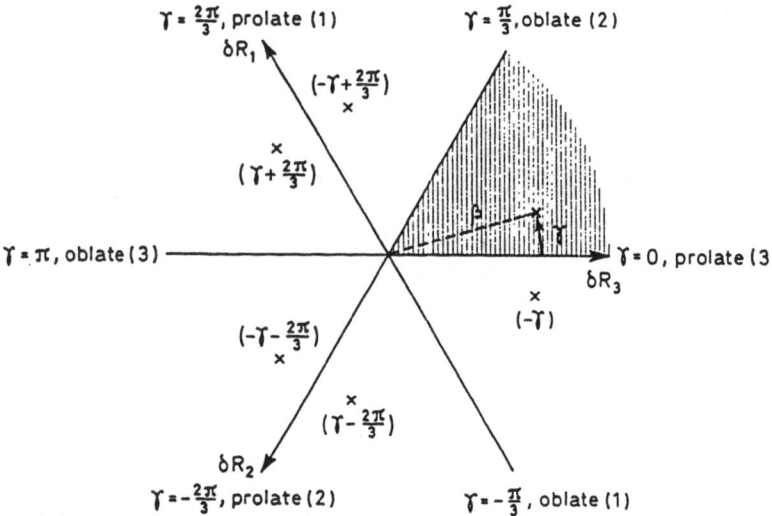

Figure 6.5: Symmetries in the (β_H, γ_H)-plane. The projections on the three axes are proportional to the increments in the principal radii of the shape. Points on the axes correspond to axially symmetric shapes. The six different points obtained by reflection represent the same shapes (from [Bohr 75]).

The relative rotational energy reads [Bohr 69]

$$B_{\text{rot}} = 1 - \alpha_H \cos\left(\gamma_H + \frac{2}{3}\pi\right) - \frac{1}{2}\alpha_H^2 \cos^2\left(\gamma_H + \frac{2}{3}\pi\right) \cdots \qquad (6.82)$$

In spherical coordinates, Eq. (6.77) converts to

$$R(\theta, \varphi) = R_0 \left[\frac{\sin^2\theta\cos^2\varphi}{a^2} + \frac{\sin^2\theta\sin^2\varphi}{b^2} + \frac{\cos^2\theta}{c^2}\right]^{-1/2}. \qquad (6.83)$$

If large ellipsoidal deformations are involved, the exponentials in Eq. (6.79) become inconvenient. A better choice is to introduce the eccentricities [Remaud 78], [Remaud 81]

$$e_1^2 = 1 - (c/a)^2$$
$$e_2^2 = 1 - (b/a)^2$$
$$e_3^2 = 1 - (c/b)^2 \,, \tag{6.84}$$

where, by definition

$$(1 - e_1^2) = (1 - e_2^2)(1 - e_3^2) \,. \tag{6.85}$$

In addition, volume conservation (6.78) must be employed. The axes are chosen in such a way that

$$a \geq b \geq c$$
$$e_1 \geq e_2, e_3 \,. \tag{6.86}$$

Conversion from (β_H, γ_H) to (e_1, e_3) is shown in Fig. 6.6.

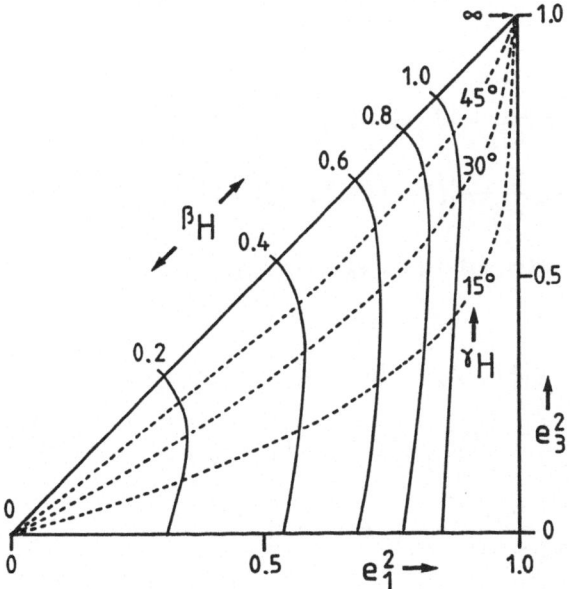

Figure 6.6: Relation between the coordinates (e_1, e_3) and (β_H, γ_H). Full (dotted) lines correspond to constant β_H (γ_H) (from [Remaud 78]).

A few geometrical quantities have been derived in [Remaud 81], cf. also [Carlson 61b], [Leander 74]. With the notation

$$\sin \psi = e_1$$
$$k_2 = e_2/e_1$$
$$k_3 = e_3/e_1 \tag{6.87}$$

and the incomplete elliptic integrals $F(\psi, k)$, $E(\psi, k)$ the relative surface, Coulomb and curvature energies are given in terms of the dimensionless axes $(a, b, c) \to (a, b, c)/R_0$,

$$B_{\text{surf}} = \frac{ab}{2}\left[\frac{1-e_1^2}{e_1}F(\psi,k_3) + e_1 E(\psi,k_3) + c^3\right]$$

$$B_{\text{Coul}} = \frac{bc}{e_1}F(\psi,k_2)$$

$$B_{\text{curv}} = \frac{bc}{2a}\left[1 + \frac{a_1^3}{e_1}((1-e_1^2)F(\psi,k_2) + e_1^2 E(\psi,k_2))\right]. \tag{6.88}$$

The redistribution energies B_{red} and B_{sr1} (B_{sr2} has not been calculated) are obtained from

$$B_{\text{red}} = \frac{3}{2}\left(5\sum_{i=1}^{3}B_i^2 - B_{\text{Coul}}^2\right)$$

$$B_{\text{sr1}} = \frac{9}{64}\left(5\sum_{i=1}^{3}A_iB_i + B_{\text{Coul}}B_{\text{surf}}\right)^2, \tag{6.89}$$

where

$$A_1 = \frac{c}{2a^2 e_1^3(1-k_3^2)}\left[(1-e_1^2)E(\psi,k_3) + (e_1^2-e_3^2)F(\psi,k_3) - \frac{e_1}{a^3}\right]$$

$$A_2 = \frac{c}{2b^2 e_1^3 k_3^2(1-k_3^2)}\left[(1-k_3^2)F(\psi,k_3) - (1-e_3^2)E(\psi,k_3) + \frac{e_1 k_3^2}{a^3}\right]$$

$$A_3 = \frac{1}{2ce_1^3 k_3^2}\left[E(\psi,k_3) - (1-e_3^2)F(\psi,k_3)\right] \tag{6.90}$$

$$B_1 = \frac{bc}{e_1^3 k_2^2}\left[F(\psi,k_2) - E(\psi,k_2)\right]$$

$$B_2 = \frac{bc(1-k_2^2 e_1^2)}{e_1^3 k_2^2(1-k_2^2)}\left[E(\psi,k_2) - (1-k_2^2)F(\psi,k_2) - \frac{ce_1 k_2^2}{b}\right]$$

$$B_3 = \frac{bc(1-e_1^2)}{e_1^3(1-k_2^2)}\left[\frac{be_1}{c} - E(\psi,k_2)\right]. \tag{6.91}$$

6.5.2 Bohr Parameterization

Bohr's parameters [Bohr 52], [Bohr 75] β_B and γ_B are defined by the shape

$$R(\theta,\varphi)$$
$$= R_0\left[1 + \beta_B\left(\cos\gamma_B Y_{20}(\theta) + \frac{\sin\gamma_B}{\sqrt{2}}(Y_{22}(\theta,\varphi) + Y_{22}^*(\theta,\varphi))\right)\right]$$
$$= R_0\left[1 + \sqrt{\frac{5}{16\pi}}\beta_B(\cos\gamma_B(3\cos^2\theta - 1) + \sqrt{3}\sin\gamma_B\sin^2\theta\cos2\varphi)\right]. \tag{6.92}$$

Other notation

$$\alpha_B = \sqrt{5/4\pi}\,\beta_B. \tag{6.93}$$

Note that this β_B differs from the one of Sect. 6.2.4 in the respect that (6.92) does not account for volume conservation, i.e. β_B should only be used linearly. However, volume conservation can be employed [Kaniowska 76] by replacing R_0 by

$$R_0(\beta_B, \gamma_B) = R_0 \left(1 - \frac{3}{4}\alpha_B^2 + \frac{1}{4}\alpha_B^3 \cos 3\gamma_B\right)^{-1/3} . \tag{6.94}$$

Here γ_B can also be restricted to $0 \le \gamma_B < \pi/3$ and Table 6.4 also holds for (β_B, γ_B). The semiaxes are

$$
\begin{aligned}
a &= R_0 \left(1 + \alpha_B \cos(\gamma_B - \frac{2\pi}{3})\right) \\
b &= R_0 \left(1 + \alpha_B \cos(\gamma_B + \frac{2\pi}{3})\right) \\
c &= R_0 \left(1 + \alpha_B \cos \gamma_B\right) ,
\end{aligned} \tag{6.95}
$$

so that in first order Bohr's parameters β_B, γ_B are equal to Hill and Wheeler's parameter β_H, γ_H of Sect. 6.5.1 . In first order of the small quantities the various parameters are related by

$$
\begin{aligned}
\frac{1}{3}e^2 &= \frac{2}{3}\varepsilon \\
&= \alpha_2 = \alpha_H = \alpha_B \\
&= \sqrt{\frac{5}{4\pi}}\beta_B = \sqrt{\frac{5}{4\pi}}\beta_H = \sqrt{\frac{5}{4\pi}}\beta \\
\gamma_B &= \gamma_H = \gamma.
\end{aligned} \tag{6.96}
$$

Let $(x, y, z) = (1, 2, 3)$ then the irrotational moments of inertia in units of the sphere (1.28) are [Ring 80]

$$\mathcal{J}_k = 1 + \alpha_B \cos\left(\gamma_B - k\frac{2\pi}{3}\right) \quad , k = 1, 2, 3 . \tag{6.97}$$

The hydrodynamic mass parameters (1.83) can be given exactly [Kaniowska 76],

$$
\begin{aligned}
B_{\beta_B \beta_B} &= \frac{3}{8\pi}a^2 \left(1 + \alpha_B^2 a^3 b_1 + \frac{1}{2}\alpha_B^2 a^5 b_1^2 b_2\right) \\
B_{\beta_B \gamma_B} &= \frac{3}{32\pi}\alpha_B^3 a^5 \left(1 + a^3 b_1 b_2\right) \sin 3\gamma_B \\
B_{\gamma_B \gamma_B} &= \frac{3}{8\pi}a^2 \left(1 + \frac{1}{8}\alpha_B^4 a^6 b_2 \sin^2 3\gamma_B\right) \\
B_{\omega_k} &= \frac{3}{8\pi}a^2 \frac{(1 - \frac{1}{2}\alpha_B \cos \gamma_k)^2}{b_2 - \alpha_B \cos \gamma_k - \frac{1}{4}\alpha_B^2 \cos 2\gamma_k} ,
\end{aligned} \tag{6.98}
$$

where

$$
\begin{aligned}
\gamma_k &= \gamma_B - k\frac{2}{3}\pi \quad k = 1, 2, 3 \\
a &= R_0(\beta_B, \gamma_B)/R_0
\end{aligned}
$$

67

$$b_1 = 1 - \frac{1}{2}\alpha_B \cos 3\gamma_B$$

$$b_2 = 1 + \frac{1}{2}\alpha_B^2 . \tag{6.99}$$

For small deformations they simplify to

$$B_{\beta_B \beta_B} = \frac{3}{8\pi}\left(1 + \frac{3}{2}\alpha_B^2 \cdots\right)$$

$$B_{\beta_B \gamma_B} = \frac{3}{32\pi}\alpha_B^3 \sin 3\gamma_B \left(1 - \frac{1}{4}\alpha_B \cos 3\gamma_B \cdots\right)$$

$$B_{\gamma_B \gamma_B} = \frac{3}{8\pi}\left(1 + \frac{1}{2}\alpha_B^2 \cdots\right)$$

$$B_{\omega_k} = \frac{3}{8\pi}\left[1 + \frac{1}{2}\alpha_B^2\left(1 - \frac{3}{2}\sin^2 \gamma_k\right)\right] . \tag{6.100}$$

The relative energies read

$$B_{\text{surf}} = 1 + \frac{2}{5}\alpha_B^2 \cdots$$

$$B_{\text{Coul}} = 1 - \frac{1}{5}\alpha_B^2 \cdots$$

$$B_{\text{rot}} = 1 + \alpha_B \cos\left(\gamma_B + \frac{2}{3}\pi\right) \cdots \tag{6.101}$$

For a rotating uniformly charged liquid drop the equilibrium configuration for $z \ll 1$ is an oblate shape described by

$$\alpha_B^0 = \frac{5}{4}\frac{z}{y} \cdots$$

$$\gamma_B^0 = \frac{\pi}{3} , \tag{6.102}$$

where $y = 1 - x$ and x is the fissility of Eq. (1.75) and z is the rotational parameter of Eq. (1.78). For larger angular momenta one has

$$\alpha_B^0 = \frac{7}{6}\left(\sqrt{1 + \frac{15z}{7y^2}} - 1\right)$$

$$\gamma_B^0 = \frac{\pi}{3}. \tag{6.103}$$

In order to have consistency with the parameter β of Sect. 6.2.4 , i.e. volume conservation up to quadratic terms, one defines

$$R(\theta, \varphi) = R_0\left[1 + \beta\left(\cos\gamma Y_{20}(\theta) + \frac{\sin\gamma}{\sqrt{2}}\left(Y_{22}(\theta, \varphi) + Y_{22}^*(\theta, \varphi)\right)\right) - \frac{\beta^2}{4\pi}\right]. \tag{6.104}$$

Then [Remaud 78]

$$\beta = \sqrt{\frac{4\pi}{5}}\alpha_H\left[1 - \frac{1}{14}\alpha_H^2 \cos 3\gamma_H\right] \cdots \tag{6.105}$$

$$\tan\gamma = \tan\gamma_H \left[1 + \frac{\alpha_H}{7}\frac{\sin 3\gamma_H}{\sin 2\gamma_H}\right]\cdots \tag{6.106}$$

6.5.3 Higher Order Triaxial Shapes

Closer approximation to ellipsoids is reached by including hexadecapole deformations [Carlson 61b], [Bohr 75], [Remaud 78],

$$R(\theta,\varphi) = \sum_{\lambda=0}^{4}\sum_{\mu=-\lambda}^{\lambda} a_{\lambda\mu}Y_{\lambda\mu}(\theta,\varphi)\,, \tag{6.107}$$

where, for reality of the surface

$$\begin{aligned}
a_{\lambda\mu} &= 0 && ,\ \lambda \text{ or } \mu \text{ odd}\\
a_{\lambda\mu} &= a_{\lambda,-\mu} && ,\ \lambda \text{ and } \mu \text{ even}\,.
\end{aligned} \tag{6.108}$$

Up to second order in the parameters one gets [Carlson 61b]

$$\begin{aligned}
a_{00} &= \sqrt{4\pi}\left(1 - \frac{1}{5}\alpha_H^2\right)\\[4pt]
a_{20} &= \sqrt{\frac{4\pi}{5}}\left[\alpha_H\cos\gamma_H - \frac{1}{14}\alpha_H^2\cos^2\gamma_H\right]\\[4pt]
a_{22} &= \sqrt{\frac{2\pi}{5}}\left[\alpha_H\sin\gamma_H + \frac{1}{14}\alpha_H^2\sin^2\gamma_H\right]\\[4pt]
a_{40} &= \sqrt{\pi}\,\frac{3}{35}\,\alpha_H^2\left[1 + 5\cos^2\gamma_H\right]\\[4pt]
a_{42} &= \sqrt{30\pi}\,\frac{3}{70}\,\alpha_H^2\sin^2\gamma_H\\[4pt]
a_{44} &= \sqrt{70\pi}\,\frac{3}{70}\,\alpha_H^2\sin^2\gamma_H\,.
\end{aligned} \tag{6.109}$$

A generalization of Bohr's parameterization to include γ-type hexadecapole deformations is provided by [Rohozinski 81]. First one defines the shape by

$$R(\theta,\phi) = R_0\left(1 + \sum_{\lambda\mu} a_{\lambda\mu}Y_{\lambda\mu}(\theta,\phi)\right) \tag{6.110}$$

$$= R_0\Big[1 + a_{20}Y_{20}(\theta,\phi) + a_{22}\left(Y_{22}(\theta,\phi) + Y_{2-2}(\theta,\phi)\right) + a_{40}Y_{40}(\theta,\phi)$$

$$+ a_{42}\left(Y_{42}(\theta,\phi) + Y_{4-2}(\theta,\phi)\right) + a_{44}\left(Y_{44}(\theta,\phi) + Y_{4-4}(\theta,\phi)\right)\Big]\,. \tag{6.111}$$

Then one expresses the coefficients a_{20}, a_{22} using Bohr's quadrupole deformation parameters $\beta_2 = \beta_B, \gamma_2 = \gamma_B$ of Eq. (6.92),

$$\begin{aligned}
a_{20} &= \beta_2\cos\gamma_2\\[4pt]
a_{22} &= \frac{\beta_2}{\sqrt{2}}\sin\gamma_2
\end{aligned} \tag{6.112}$$

and the coefficients a_{40}, a_{42}, a_{44} by the new hexadecapole deformation parameters $\beta_4, \delta_4, \gamma_4$,

$$a_{40} = \beta_4 \left(\sqrt{\frac{7}{12}} \cos \delta_4 + \sqrt{\frac{5}{12}} \sin \delta_4 \cos \gamma_4 \right)$$

$$a_{42} = -\frac{1}{\sqrt{2}} \beta_4 \sin \delta_4 \sin \gamma_4$$

$$a_{44} = \frac{1}{\sqrt{2}} \beta_4 \left(\sqrt{\frac{5}{12}} \cos \delta_4 - \sqrt{\frac{7}{12}} \sin \delta_4 \cos \gamma_4 \right) . \quad (6.113)$$

The inverse transformation is given by

$$\beta_2 = \sqrt{a_{20}^2 + 2a_{22}^2}$$

$$\tan \gamma_2 = \sqrt{2} \frac{a_{22}}{a_{20}}$$

$$\beta_4 = \sqrt{a_4^2 + b_4^2 + c_4^2}$$

$$\tan \gamma_4 = \frac{c_4}{b_4}$$

$$\sin \delta_4 = \sqrt{\frac{b_4^2 + c_4^2}{a_4^2 + b_4^2 + c_4^2}}, \quad (6.114)$$

where

$$a_4 = \sqrt{\frac{7}{12}} a_{40} + \sqrt{\frac{5}{6}} a_{44}$$

$$b_4 = \sqrt{\frac{5}{12}} a_{40} - \sqrt{\frac{7}{6}} a_{44}$$

$$c_4 = -\sqrt{2} a_{42} . \quad (6.115)$$

It has been shown by [Rohozinski 81] that there is a one to one correspondence between the shape (6.111) and the parameters $\beta_2, \gamma_2, \beta_4, \gamma_4, \delta_4$ if the latter are inside the region

$$\beta_2 \geq 0$$
$$0 \leq \gamma_2 \leq \pi/3$$
$$\beta_4 \geq 0$$
$$0 \leq \gamma_4 \leq \pi/3$$
$$0 \leq \delta_4 \leq \pi . \quad (6.116)$$

In other words, the new parameters describe the shape (6.111) in an invariant way not depending on the designation of the intrinsic axes or on the choice of the positive direction. Axial symmetry of the hexadecapole shape is obtained for $\delta_4 = \delta_4^0, \gamma_4 = 0$ (with respect to the z-axis) and for $\delta_4 = \pi - \delta_4^0, \gamma_4 = \pi/3$ (with respect to the y-axis), where $\cos \delta_4^0 = \sqrt{7/12}$. Simultaneous axial

70

symmetry (with respect to the z-axis as well as with respect to the y-axis) of both the quadrupole and the hexadecapole shapes may be achieved by relating the quadrupole and hexadecapole deformation parameters in a proper way, e.g. by the relation $\cos \delta_4 = \delta_4^0 \cos 3\gamma_2$, $\gamma_4 = \gamma_2$.

Triaxiality has been incorporated into the Nilsson potential by [Larsson 73]. In the stretched coordinates defined in Eq. (6.122) it reads

$$
\begin{aligned}
V_{\text{osc}}(\boldsymbol{r}_t) &= \frac{\hbar}{2}(x_t^2\omega_x + y_t^2\omega_y + z_t^2\omega_z) \\
&= \frac{1}{2}\hbar\omega_0(\varepsilon, \gamma_\varepsilon)r_t^2 \left\{ 1 - \frac{2}{3}\varepsilon \sqrt{\frac{4\pi}{5}} \Big[\cos \gamma_\varepsilon Y_{20}(\theta_t, \phi_t) \right. \\
&\qquad \left. - \frac{\sin\gamma_\varepsilon}{\sqrt{2}} \big(Y_{22}(\theta_t, \phi_t) + Y_{2-2}(\theta_t, \phi_t) \big) \Big] \right\},
\end{aligned}
\tag{6.117}
$$

so that the frequencies in the different directions become

$$
\begin{aligned}
\omega_x &= \omega_0(\varepsilon, \gamma_\varepsilon)\left[1 - \frac{2}{3}\varepsilon \cos\left(\gamma_\varepsilon + \frac{2\pi}{3} \right) \right] \\
\omega_y &= \omega_0(\varepsilon, \gamma_\varepsilon)\left[1 - \frac{2}{3}\varepsilon \cos\left(\gamma_\varepsilon - \frac{2\pi}{3} \right) \right] \\
\omega_z &= \omega_0(\varepsilon, \gamma_\varepsilon)\left[1 - \frac{2}{3}\varepsilon \cos\gamma_\varepsilon \right].
\end{aligned}
\tag{6.118}
$$

Unfortunately, the sign of γ_ε here is reversed as compared to Eq. (6.95) (Lund convention). Later on, however, the same sign has been used again by [Rohozinski 81].

By introducing the new parameters

$$
\begin{aligned}
\varepsilon_2 &= \varepsilon \cos\gamma_\varepsilon \\
\varepsilon_{22} &= \frac{2}{\sqrt{3}}\varepsilon \sin\gamma_\varepsilon
\end{aligned}
\tag{6.119}
$$

the frequencies become

$$
\begin{aligned}
\omega_x &= \omega_0(\varepsilon_2, \varepsilon_{22})\left(1 + \frac{1}{3}\varepsilon_2 + \frac{1}{2}\varepsilon_{22} \right) \\
\omega_y &= \omega_0(\varepsilon_2, \varepsilon_{22})\left(1 + \frac{1}{3}\varepsilon_2 - \frac{1}{2}\varepsilon_{22} \right) \\
\omega_z &= \omega_0(\varepsilon_2, \varepsilon_{22})\left(1 - \frac{2}{3}\varepsilon_2 \right)
\end{aligned}
\tag{6.120}
$$

and volume conservation demands

$$
\omega_x\omega_y\omega_z = \overset{\circ}{\omega}_0^{\,3}.
\tag{6.121}
$$

The transformation between the normal and the stretched coordinates reads

$$
\begin{aligned}
(r_t)_i &= r_i\sqrt{\frac{M\omega_i}{\hbar}} \quad , i = x, y, z \\
\cos\theta_t &= \cos\theta\sqrt{\frac{1 - \frac{2}{3}\varepsilon_2}{1 + \frac{1}{3}\varepsilon_2 - \varepsilon_2\cos^2\gamma_\varepsilon + \frac{1}{2}\varepsilon_{22}\sin^2\theta\cos 2\varphi}}
\end{aligned}
$$

71

$$\tan\varphi_t = \tan\varphi\sqrt{\frac{1+\frac{1}{3}\varepsilon_2-\frac{1}{2}\varepsilon_{22}}{1+\frac{1}{3}\varepsilon_2+\frac{1}{2}\varepsilon_{22}}} \,, \tag{6.122}$$

so that the potential becomes

$$V_{\text{osc}}(r) = \frac{1}{2}M(\omega_x^2 x^2 + \omega_y^2 y^2 + \omega_z^2 z^2) \,. \tag{6.123}$$

The hydrodynamic mass parameter (1.8) with respect to γ_ε becomes [Möller 81]

$$B_{\gamma_\varepsilon} = \frac{2}{15}\left(\frac{1-\frac{2}{3}\varepsilon}{1+\frac{1}{3}\varepsilon}\right)^{2/3}\left(\log\frac{1+\frac{1}{3}\varepsilon}{1-\frac{2}{3}\varepsilon}\right)^2 \tag{6.124}$$

and the mass parameter with respect to ε is given in Eq. (5.27).

Chapter 7

Large Deformations

7.1 Arbitrary Shapes

7.1.1 Cylindrical Coordinates

7.1.1.1 General Formulae

Let $z_0 = R_0\zeta_0$ be half the length of the shape and $z = R_0\zeta$, $\varrho = \varrho(\zeta)$ be the dimensionless z-coordinate and the axially symmetric shape function in cylindrical coordinates, respectively, then the relevant quantities are calculated from multiple integrals (if not otherwise noted, integrals run from $-\zeta_0$ to $+\zeta_0$). In the natural units of Chap. 1,

$$Q_\ell = \frac{4\pi}{2\ell+3} \int d\zeta \left(\varrho^2 + \zeta^2\right)^{\ell/2+1} [P_\ell(x) - P_{\ell+2}(x)]$$

$$\text{where } x = \frac{\zeta}{\sqrt{\varrho^2 + \zeta^2}} \tag{7.1}$$

$$Q = \frac{\pi}{2} \int d\zeta \, \varrho^2(4\zeta^2 - \varrho^2)$$

$$Q_4 = \frac{\pi}{4} \int d\zeta \, \varrho^2(8\zeta^4 - 12\zeta^2\varrho^2 + \varrho^4) \tag{7.2}$$

$$r_{rms}^2 = \frac{3}{8} \int d\zeta \, \varrho^2(2\zeta^2 + \varrho^2)$$

$$\overline{(\delta r)^2} = \frac{1}{2} \int d\zeta \, \varrho \frac{\varrho - \zeta\varrho'}{(\varrho^2 + \zeta^2)^{3/2}} \left(\sqrt{\varrho^2 + \zeta^2} - 1\right)^2 \tag{7.3}$$

$$\mathcal{J}_\parallel = \frac{15}{16} \int d\zeta \, \varrho^4$$

$$\mathcal{J}_\perp = \frac{15}{32} \int d\zeta \, \varrho^2(4\zeta^2 + \varrho^2) \tag{7.4}$$

$$B_{surf} = \frac{1}{2} \int d\zeta \, \varrho\sqrt{1 + \varrho'^2}$$

$$\text{with } \varrho' = d\varrho(\zeta)/d\zeta . \tag{7.5}$$

Eq. (7.1) is from [Pashkevich 83] and the following equations are from [Hasse 71]. The relative Coulomb energy is due to [Gray 19], [Lawrence 65]

$$B_{Coul} = \frac{15}{4\pi} \int d\zeta \int\limits_{-\zeta_0}^{\zeta} d\bar{\zeta} \int\limits_0^\pi d\phi \, \frac{\varrho^2\bar{\varrho}^2 \sin^2\phi}{\zeta - \bar{\zeta} + f}$$

$$f = \sqrt{(\zeta - \overline{\zeta})^2 + \varrho^2 + \overline{\varrho}^2 - 2\varrho\overline{\varrho}\cos\phi}$$

$$\text{with}\quad \overline{\varrho} = \varrho(\overline{\zeta}) \tag{7.6}$$

$$B_{\text{KN}} = \frac{2}{3}\sigma + \frac{1}{\sigma^2} - \left(1 + \frac{1}{\sigma}\right)^2 e^{-2\sigma} - g$$

$$\text{with } g = \frac{1}{4\pi}\int d\zeta \int d\overline{\zeta} \int_0^{2\pi} d\phi \frac{\varrho\overline{\varrho}}{f^4}\left[\varrho - \overline{\varrho}\cos\phi - \varrho'(\zeta - \overline{\zeta})\right]$$

$$\times \left[\overline{\varrho} - \varrho\cos\phi - \overline{\varrho}'(\overline{\zeta} - \zeta)\right]\left[\sigma f + (2 + \sigma f)e^{-\sigma f} - 2\right] \tag{7.7}$$

$$B_{\text{YE}} = \frac{1}{4\pi}\int d\zeta \int d\overline{\zeta} \int_0^{2\pi} d\phi \frac{\varrho\overline{\varrho}}{f^4}\left[\varrho - \overline{\varrho}\cos\phi - \varrho'(\zeta - \overline{\zeta})\right]$$

$$\times \left[\overline{\varrho} - \varrho\cos\phi - \overline{\varrho}'(\overline{\zeta} - \zeta)\right]\left[2 - ((\sigma f)^2 + 2\sigma f + 2)e^{-\sigma f}\right] \tag{7.8}$$

$$B_{\text{curv}} = \frac{1}{4}\int d\zeta \frac{1 + \varrho'^2 - \varrho\varrho''}{1 + \varrho'^2}$$

$$\text{with}\quad \varrho'' = d^2\varrho(\zeta)/d\zeta^2 . \tag{7.9}$$

$$B_{\text{red}} = \frac{175}{3}\left(\overline{\Phi^2} - \overline{\Phi}^2\right)$$

$$\overline{\Phi} = \frac{6}{5}B_{\text{Coul}}$$

$$\overline{\Phi^2} = \frac{3}{2}\int d\zeta \int_0^{\varrho(\zeta)} d\varrho\varrho\, \Phi^2(\varrho,\zeta) \tag{7.10}$$

$$\Phi(\varrho,\zeta) = \frac{3}{4\pi}\int d\overline{\zeta}\frac{k}{\sqrt{\varrho\overline{\varrho}}}\left\{[\varrho\overline{\varrho} + \overline{\varrho}^2 + (\zeta - \overline{\zeta})\varrho\overline{\varrho}']K(k^2) - 2\varrho\overline{\varrho}\, D(k^2)\right\}$$

with $K(k^2)$ and $D(k^2) = [K(k^2) - E(k^2)]/k^2$ being complete elliptic integrals of the argument

$$k^2 = \frac{4\varrho\overline{\varrho}}{(\zeta - \overline{\zeta})^2 + (\varrho + \overline{\varrho})^2} \tag{7.11}$$

$$B_{\text{sr1}} = (5\overline{\Phi}_s)^2 - 60\overline{\Phi}_s B_{\text{Coul}}B_{\text{surf}} + (6B_{\text{Coul}}B_{\text{surf}})^2$$

$$\overline{\Phi}_s = \frac{1}{2}\int d\zeta\, \varrho\sqrt{1 + \varrho'^2}\, \Phi_s(\zeta)$$

$$\text{with } \Phi_s(\zeta) = \Phi(\varrho(\zeta),\zeta) \tag{7.12}$$

$$B_{\text{sr2}} = 25\overline{\Phi_s^2} - 60\overline{\Phi}_s B_{\text{Coul}} + 36B_{\text{Coul}}^2 B_{\text{surf}}$$

$$\overline{\Phi_s^2} = \frac{1}{2}\int d\zeta\, \varrho\sqrt{1 + \varrho'^2}\, \Phi_s^2(\zeta) . \tag{7.13}$$

The relative Coulomb energy can also be obtained by direct integration over the Coulomb potential [Nix 69] ,

$$B_{\text{Coul}} = \int d\zeta \; \Phi_s(\zeta)(\varrho^2 - \zeta \varrho \varrho').$$ (7.14)

For a discussion on the numerical evaluation of the Coulomb energy with the two methods, see [Davies 75].

Eqs. (7.14) and (7.11) can also be used even if the shape has an infinite slope, for instance in the calculation of the Coulomb-self energy of a hemisphere, cf. Eq. (7.98). In this case the expression $\varrho' d\zeta$ is to be replaced by $d\varrho$ and the integration over $d\varrho$ extends over the cross section of the cut at fixed ζ. The Coulomb-interaction energy of two bodies with sharp cuts, for instance the one of two hemispheres, can be calculated in the same way by restricting the integration in the Coulomb potential to one body and the integration in the Coulomb energy over the other. However, since the second part of Eq. (7.14) is not translational invariant, one always has to sum up the interaction energies of both objects.

Another fast numerical method is due to [Beringer 63], where the Coulomb energy of a cylindrically symmetric shape is computed by summing over (altogether 210) tabulated values of the Coulomb interaction energy between thin cylindrical disks of different diameters and different separations.

The hydrodynamic mass parameters and two-body viscosity and one-body dissipation coefficients of Sect. 1.8 can be evaluated in Werner-Wheeler approximation, cf. [Davies 76b]. For a given shape function $\varrho_s(\zeta, \alpha)$ and a given set of dimensionless deformation parameters $\{\alpha\}$ first calculate the auxiliary functions

$$A_i(\zeta, \alpha) = \varrho_s^{-2} \frac{\partial}{\partial \alpha_i} \int_{\zeta}^{\zeta_{\text{max}}} d\bar{\zeta} \varrho_s^2(\bar{\zeta}, \alpha) ,$$ (7.15)

where ζ_{max} is the larger zero of ϱ_s. Then mass parameters and viscosity coefficients read

$$B_{ij} = \frac{3}{4} \int_{\zeta_{\text{min}}}^{\zeta_{\text{max}}} d\zeta \varrho_s^2 (A_i A_j + \frac{1}{8} \varrho_s^2 A_i' A_j')$$

$$Z_{ij} = \frac{1}{4} \int_{\zeta_{\text{min}}}^{\zeta_{\text{max}}} d\zeta \varrho_s^2 (3 A_i' A_j' + \frac{1}{8} \varrho_s^2 A_i'' A_j'') ,$$ (7.16)

where ζ_{min} is the smaller zero of ϱ_s. The one-body dissipation coefficients, on the other hand, are given exactly by, cf. [Sierk 80],

$$D_{ij} = \frac{1}{2} \int_{\zeta_{\text{min}}}^{\zeta_{\text{max}}} d\zeta \varrho_s \left(1 + \varrho_s'^2\right)^{-1/2} \frac{\partial \varrho_s}{\partial \alpha_i} \frac{\partial \varrho_s}{\partial \alpha_j}.$$ (7.17)

7.1.1.2 Variational Method

In general, the LDM energy of a nucleus for given deformation is calculated with an underlying family of shapes which is always restrictive. However, by making the energy functional stationary with respect to shape changes under the constraints of volume conservation (fixed by the Lagrange parameter λ_0) and of an arbitrary deformation (described by the function $f(\zeta, \varrho)$ and fixed by the Lagrange parameter λ_1) in case of only surface and Coulomb energies one obtains the following integro-differential equation [Bohr 39], [Strutinsky 62], [Strutinsky 63]

$$\varrho\varrho'' = 1 + \varrho'^2 - \varrho[\lambda_0 + \lambda_1 f(\zeta, \varrho) + 10x\,\Phi_s(\zeta)][1 + \varrho'^2]^{3/2} . \tag{7.18}$$

For instance $f(\zeta, \varrho) =\mid \zeta \mid$ fixes the distance between the point $\zeta = 0$ of the symmetry axis and the center-of-mass of the part of the nucleus situated on one side side of the meridian plane. Eq. (7.18) is of integro-differential type because $\Phi_s(\zeta)$, the Coulomb potential at the surface, see Eq. (7.12), depends on the shape function itself. The solution of (7.18), however, can be iterated easily with the starting value $\Phi_s(\zeta) = 0$.

7.1.2 Spherical Coordinates

Let $R(\theta) = R_0 r(\theta)$ be the axially symmetric shape function in spherical coordinates then the relevant quantities are calculated from multiple integrals, where $\mu = \cos\theta$, $-1 \le \mu \le 1$, $0 \le \theta \le \pi$, $P_k = P_k(\mu)$. In the natural units of Chap. 1,

$$Q = \frac{4}{5}\pi \int d\mu\, r^5 P_2$$
$$Q_4 = \frac{4}{7}\pi \int d\mu\, r^7 P_4 \tag{7.19}$$

$$r^2_{\text{rms}} = \frac{3}{10} \int d\mu\, r^5$$
$$\overline{(\delta r)^2} = \frac{1}{2} \int d\mu\, (r - 1)^2 \tag{7.20}$$

$$\mathcal{J}_\| = \frac{1}{2} \int d\mu\, r^5 (1 - P_2)$$
$$\mathcal{J}_\perp = \frac{1}{4} \int d\mu\, r^5 (2 + P_2) \tag{7.21}$$

$$B_{\text{surf}} = \int d\mu\, r\sqrt{r^2 + r'^2}$$
$$\text{with} \quad r' = dr(\theta)/d\theta \tag{7.22}$$

$$B_{\text{Coul}} = \frac{1}{2} \int d\theta\, r^3\, \Phi_s(\theta)$$
$$\Phi(r,\theta) = \frac{3}{2} \int d\overline{\mu} \left\{ r^2 P_2 \overline{P}_2 \left(\frac{1}{5} + \log\frac{\overline{r}}{r} \right) \right.$$

$$+ \sum_{k \neq 2} P_k \overline{P}_k \left(\frac{r^2}{k+3} + \frac{r^2 - r^k \, \overline{r}^{2-k}}{k-2} \right) \bigg\}$$

$$\text{with} \quad \overline{r} = r(\overline{\theta}), \; \overline{P}_k = P_k(\overline{\mu})$$
$$\text{and} \quad \Phi_s(\theta) = \Phi(r(\theta), \theta) \tag{7.23}$$

$$B_{\text{curv}} = \frac{1}{4} \int d\theta \left\{ r \sin\theta \frac{2r^2 + 3r'^2 - r \, r''}{r^2 + r'^2} - r' \cos\theta \right\}$$
$$\text{with} \quad r'' = d^2 r(\theta)/d\theta^2 \tag{7.24}$$

$$B_{\text{red}} = \frac{175}{3} \left(\overline{\Phi^2} - \overline{\Phi}^2 \right)$$
$$\overline{\Phi} = \frac{6}{5} B_{\text{Coul}}$$
$$\overline{\Phi^2} = \frac{3}{2} \int d\mu \sin\theta \int\limits_0^{r(\theta)} dr \, r^2 \Phi^2(r, \theta) \tag{7.25}$$

$$B_{\text{sr1}} = (5\overline{\Phi}_s)^2 - 60 \overline{\Phi}_s B_{\text{Coul}} B_{\text{surf}} + (6 B_{\text{Coul}} B_{\text{surf}})^2$$
$$\overline{\Phi}_s = \int d\mu \, r \sqrt{r^2 + r'^2} \, \Phi_s(\theta) \tag{7.26}$$

$$B_{\text{sr2}} = 25 \overline{\Phi_s^2} - 60 \overline{\Phi}_s B_{\text{Coul}} + 36 B_{\text{Coul}}^2 B_{\text{surf}})$$
$$\overline{\Phi_s^2} = \int d\mu \, r \sqrt{r^2 + r'^2} \, \Phi_s^2(\theta) \; . \tag{7.27}$$

7.2 Generalized Spheroids

7.2.1 Lawrence Shapes

The quartic
$$\varrho_s^2 = az^4 + bz^2 + c \tag{7.28}$$
has first been used by Lawrence [Lawrence 65], [Lawrence 67] to describe spherical, spheroidal, constricted, scission, and separated shapes. Here one of the parameters must be eliminated by volume conservation. A convenient method to do this and to generalize to asymmetric shapes is to write [Hasse 68a], [Hasse 68b], [Hasse 69]

$$\varrho_s^2 = R_0^3 \lambda \left(z_0^2 - (z + z_s)^2 \right) \left(z_2 \, | \, z_2 \, | + (z + z_s - z_1)^2 \right) \; . \tag{7.29}$$

Here, z_0 is half the length, z_2 is a constriction or necking-in parameter with the special cases

$$\begin{array}{lll} z_0 = R_0, & z_2 \to \infty & \text{, sphere} \\ z_0 \neq R_0, & z_2 \to \infty & \text{, spheroids} \\ z_2 = 0 & & \text{, scission} \\ z_2 < 0 & & \text{, separated shapes ,} \end{array}$$

z_1 is the asymmetry parameter with $z_1=0$ allowing for symmetric shapes. Volume conservation demands

$$\lambda^{-1} = \begin{cases} z_0^3(\frac{1}{5}z_0^2 + z_1^2 + z_2^2) & , z_2 \geq 0 \\ z_0^3(\frac{1}{5}z_0^2 + z_1^2 - z_2^2) + z_2^3(\frac{1}{5}z_2^2 + z_1^2 - z_0^2) & , z_2 \leq 0 . \end{cases} \tag{7.30}$$

Center-of-mass conservation gives

$$-z_s = \begin{cases} \frac{2}{5}\lambda z_1 z_0^5 & , z_2 \geq 0 \\ \lambda z_1 \left[\frac{2}{5}z_0^2 + z_2^3(z_0^2 - z_1^2 - \frac{3}{5}z_2^2)\right] & , z_0 \leq 0 . \end{cases} \tag{7.31}$$

The neck radius for symmetric shapes is

$$\varrho_n = R_0^{3/2} z_0 z_2 \lambda^{1/2} . \tag{7.32}$$

The dimensionless parameters

$$\zeta_i = z_i/R_0 \quad , i = 0, 1, 2, s \tag{7.33}$$

have also been employed [Hasse 69]. Fig. 7.1 shows some symmetric shapes of this parameterization. Note that the scission configuration is not given by two tangent spheres. For connected symmetric shapes ($\zeta_1 = 0$, $\zeta_2 \geq 0$) the geometrical quantities in the dimensionless coordinates read

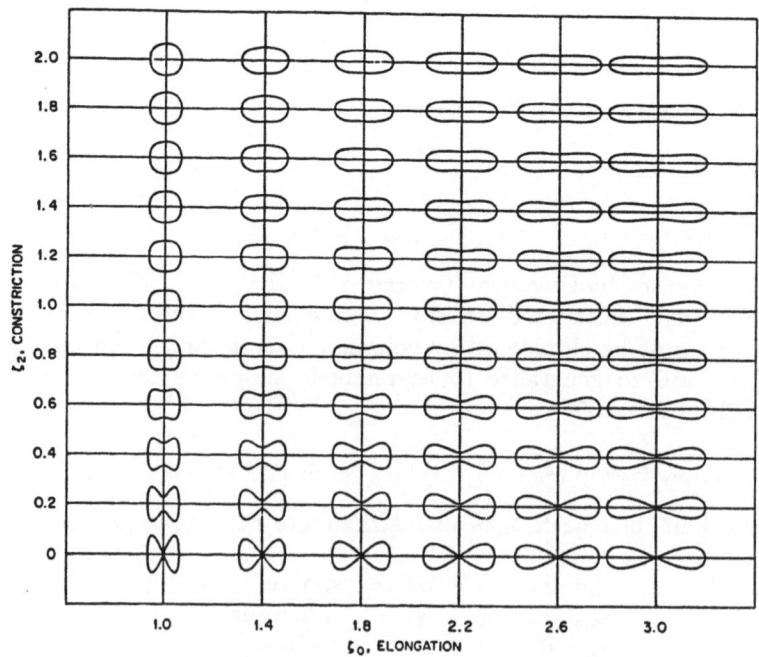

Figure 7.1: Symmetric generalized spheroids in the ($\zeta_0 - \zeta_2$)-parameterization (from [Hasse 71]).

$$\lambda^{-1} = \zeta_0^3 \left(\frac{1}{5}\zeta_0^2 + \zeta_2^2 \right) \tag{7.34}$$

$$\varrho_n = \zeta_2 \left[\zeta_0 \left(\frac{1}{5}\zeta_0^2 + \zeta_2^2 \right) \right]^{-1/2} = \zeta_0\zeta_2\sqrt{\lambda} \tag{7.35}$$

$$\begin{aligned}
Q &= \frac{8}{5}\pi\zeta_0^5\lambda \left[\frac{1}{7}\zeta_0^2 + \frac{1}{3}\zeta_2^2 - \frac{1}{3}\lambda \left(\frac{1}{21}\zeta_0^4 + \frac{2}{7}\zeta_0^2\zeta_2^2 + \zeta_2^4 \right) \right] \\
r_{\text{rms}}^2 &= \frac{1}{5}\zeta_0^5\lambda \left(\frac{3}{7}\zeta_0^2 + \zeta_2^2 + \frac{2}{21}\zeta_0^4 + \frac{4}{7}\zeta_0^2\zeta_2^2 + 2\zeta_2^4 \right)
\end{aligned} \tag{7.36}$$

$$\begin{aligned}
\mathcal{J}_\parallel &= \zeta_0^5\lambda \left(\frac{1}{21}\zeta_0^4 + \frac{2}{7}\zeta_0^2\zeta_2^2 + \zeta_2^4 \right) \\
\mathcal{J}_\perp &= \frac{1}{2}\left[\mathcal{J}_\parallel + \zeta_0^5\lambda \left(\frac{3}{7}\zeta_0^2 + \zeta_2^2 \right) \right] .
\end{aligned} \tag{7.37}$$

Other quantities are to be calculated numerically with the help of Sect. 7.1.
Various other notations are used in the literature, e.g. [Schirmer 73]

$$\begin{aligned}
\zeta_0 &= \xi_2 \\
\zeta_1 &= \xi_3\xi_2/\xi_4 \\
\zeta_2^2 &= \xi_2^2 \frac{1-\xi_4}{\xi_4}(1-\xi_3^2/\xi_4)
\end{aligned} \tag{7.38}$$

so that spheroids are obtained with $\xi_4 = 0$ and scission with $\xi_4 = 1$, or
[Sanders 72]

$$\begin{aligned}
\kappa_2 &= \zeta_0 \\
\kappa_3 &= \zeta_1 \\
\kappa_4 &= \frac{\zeta_0^2}{\zeta_0^2 + 5\zeta_2^2} = 1 - \frac{\text{neck area}}{\text{spheroidal neck area}}
\end{aligned} \tag{7.39}$$

valid only for prescission shapes. Here $\kappa_4 = 1$ gives scission and the spheroidal
neck area is the area of the corresponding spheroid, i.e. with $\zeta_2 \to \infty$.
Albrecht [Albrecht 73] used

$$\varrho_s^2 = \left(R_0^2\lambda_0^2 - \left(\frac{z}{s+1} \right)^2 \right) \left(\left[\frac{c}{s+1}\frac{z}{R_0\lambda_0} + a \right]^2 + g(c) \right), \tag{7.40}$$

where

$$g(c) = \begin{cases} 1 & , c \leq 1 \\ c(2-c) & , c \geq 1 . \end{cases} \tag{7.41}$$

$$\frac{1}{\lambda_0^3(s+1)} = \tag{7.42}$$
$$\frac{1}{5}c^2 + a^2 + g(c) + \begin{cases} 0 & , c \leq 2 \\ \left(1 - \frac{2}{c}\right)^{3/2}\left(\frac{2}{5}c(2c+1) - a^2\right) & , c \geq 2 . \end{cases}$$

Half length,

$$z_0 = \lambda_0 R_0(s+1) \,, \tag{7.43}$$

and neck radius for symmetric shapes and $c \leq 2$

$$\varrho_n = R_0 \lambda_0 \sqrt{g(c)} \,. \tag{7.44}$$

The separation parameter s is restricted to $-1 < s \leq \infty$, the constriction parameter to $0 \leq c \leq \infty$, and the asymmetry parameter to $a \geq 0$. The sphere obtains for $s = c = a = 0$, spheroids for $c = a = 0$, scission for $c = 2$ and separated shapes for $c > 2$.

7.2.2 Higher Algebraic Shapes

Higher powers than fourth order were introduced in [Trentalange 80] . Symmetric shapes read

$$\varrho_s^2(z) = R_0^2 \sum_{n=0}^{N} a_n P_n \left(\frac{z}{z_0} \right) \,, \tag{7.45}$$

where n runs over even integers and z_0 is half the length, hence

$$a_0 = - \sum_{n=2}^{N} a_n \,. \tag{7.46}$$

Volume conservation

$$z_0 = \frac{2R_0}{3a_0} \,. \tag{7.47}$$

To first order in the parameters, the connection with the harmonic parameterization of Sect. 6.2.1 is

$$a_2 = \frac{2}{3}(\alpha_2 - 1) - \frac{4}{3} \sum_{n=4}^{N} \alpha_n \tag{7.48}$$

$$a_n = 2\alpha_n \text{ for } n \geq 4 \,. \tag{7.49}$$

7.2.3 Funny Hills Parameterization

Spheroidal deformations do not suffice for ground state shapes of many nuclei. The Copenhagen group [Brack 72], [Götz 72], [Pauli 73] therefore modified the algebraic parameterization above to obtain

$$\varrho_s^2 = \begin{cases} (C^2 - z^2) \left(A + B \left(\frac{z}{C} \right)^2 + \alpha \frac{z}{C} \right) & , B \geq 0 \\[2ex] (C^2 - z^2) \left(A + \alpha \frac{z}{C} \right) e^{BCz^2/R_0^3} & , B \leq 0 \,. \end{cases} \tag{7.50}$$

Special cases:

$B = 0$,	$A = 1$	, sphere
$B = 0$,	$0 < A < 1$	, prolate spheroids
$B = 0$,	$A > 1$	, oblate spheroids
$B > 0$,	$A \geq 0$	, necked in shapes
$B > 0$,	$A < 0$	, separated shapes
$B < 0$		, lemons (diamonds) .

Volume conservation for $\alpha = 0$ results from

$$(R_0/C)^3 = \begin{cases} A + \frac{1}{5}B & , \ B \geq 0 , A \geq 0 \\[2mm] A + \frac{1}{5}B + (B + \frac{1}{5}A)(-A/B)^{3/2} & , \ B \geq 0 , A \leq 0 \end{cases}$$

$$-\frac{4B}{3A} = e^{-q} + \left(1 - \frac{1}{2q}\right)\sqrt{\pi q}\,\mathrm{erf}(\sqrt{q}) \quad , \ B \leq 0$$

$$q = -BC^3/R_0^3 , \tag{7.51}$$

where $\mathrm{erf}(x)$ is the error function. In practical use, A is eliminated by volume conservation and the parameters

$$c = C/R_0$$
$$h = \frac{1}{2}B - \frac{1}{4}(C/R_0 - 1) \tag{7.52}$$

are employed. The quantity α in (7.50) and below is an asymmetry parameter. Fig. 7.2 shows some shapes contained in the (c, h) parameterization and Fig. 7.3 shows the conversion from (β_2, β_4) of Sect. 6.2.4 to (c, h).

For connected shapes the relations of these parameters to the ones of the generalized spheroids above are

$$A = R_0^3 \lambda(z_1^2 + z_2^2)$$
$$B = R_0^3 \lambda z_0^2$$
$$C = z_0$$
$$\alpha = -2R_0^3 \lambda z_0 z_1$$
$$c = \zeta_0$$
$$h = \left[2\zeta_0\left(\frac{1}{5}\zeta_0^2 + \zeta_1^2 + \zeta_2^2\right)\right]^{-1} - \frac{1}{4}(\zeta_0 - 1)$$
$$\alpha = -2\left[\zeta_0^2\left(\frac{1}{5}\zeta_0^2 + \zeta_1^2 + \zeta_2^2\right)\right]^{-1}. \tag{7.53}$$

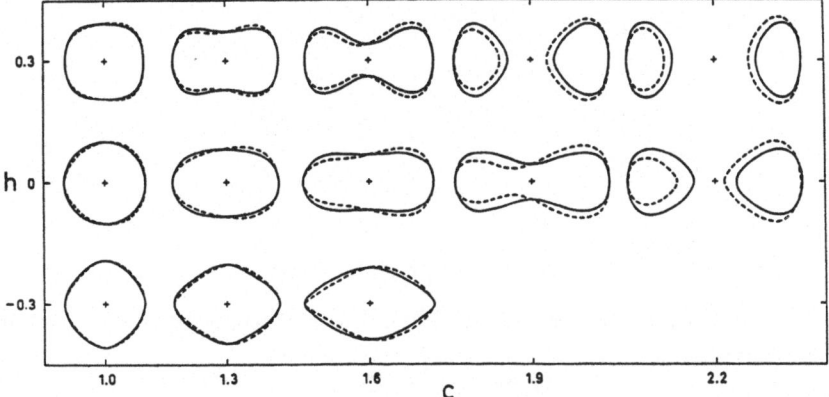

Figure 7.2: Some shapes in the (c, h)-parameterization. The full curves correspond to symmetric shapes ($\alpha = 0$) and the dotted ones to asymmetric shapes ($\alpha = 0.2$) (from [Brack 72]).

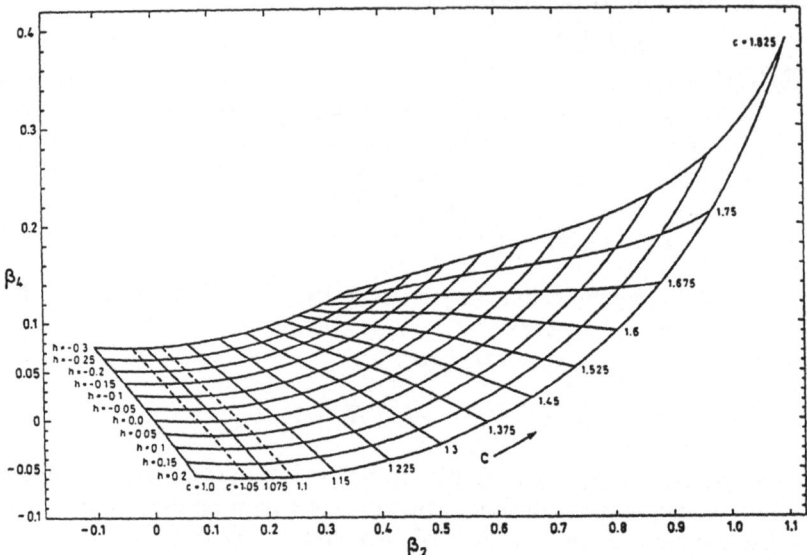

Figure 7.3: Relation between the parameters (β_2, β_4) and (c, h) (from [Brack 72]).

In the (c, h, α) parameterization the relative moments of inertia read [Saupe 87]

$$\mathcal{J}_\| = \frac{1}{c} + \frac{1}{7}c^2 f\left(\frac{1}{3}c^3 f - 1\right) + \frac{1}{7}c^5\alpha^2$$

$$\mathcal{J}_\perp = \frac{1}{2}c^2 + \frac{1}{2c} + \frac{1}{7}c^2 f\left(c^3 + \frac{1}{6}c^3 f - \frac{1}{2}\right) - \frac{1}{2}c^5\alpha^2\left(\frac{1}{5}c^3 - \frac{1}{7}\right), \quad (7.54)$$

where

$$f = \frac{4}{5}\left(2h + \frac{c-1}{2}\right). \quad (7.55)$$

The Funny-hills [Brack 72], [Damgaard 69] shell model potential, which is of Woods-Saxon form with a constant skin thickness, is constructed as follows. First rewrite Eq. (7.50) in dimensionless coordinates

$$z = C\left(u - \bar{u}\right), \qquad \varrho = Cv$$

$$v_s^2 = \begin{cases} (1 - u^2)(A + Bu^2 + \alpha u) & , B \geq 0 \\ \\ (1 - u^2)(A + \alpha u)\, \mathrm{e}^{-qu^2} & , B \leq 0 \end{cases}, \quad (7.56)$$

where

$$\bar{u} = \frac{1}{5}\alpha c^3 \quad (7.57)$$

ensures center-of-mass conservation. Then the surface constraint is defined through $(\boldsymbol{w} = (u, v))$

$$\pi(\boldsymbol{w}) = v^2 - v_s^2(u). \quad (7.58)$$

This, in turn, defines the new shape function

82

$$\Pi(\boldsymbol{w}) = \sqrt{\pi - \pi_{\min}} - \sqrt{-\pi_{\min}}, \tag{7.59}$$

where $\pi_{\min}$ is the minimum value of π for connected shapes or 0 for disconnected ones. The normal coordinate

$$\ell(\boldsymbol{r}) = \frac{\Pi(\boldsymbol{w})}{R_0 \, |\nabla \Pi(\boldsymbol{w})|} \tag{7.60}$$

then serves in the Woods-Saxon potential with strength V_0 and diffuseness a ,

$$V_{\text{WS}}(\boldsymbol{r}) = -\frac{V_0}{1 + e^{\ell(\boldsymbol{r})/a}}. \tag{7.61}$$

An alternative method of defining $\ell(\boldsymbol{r})$ can be found in [Balazs 78].

Instead of the parameter c also the parameter ϱ_{cm} has been used which is half the distance between the centers of mass of the two halves of the nucleus defined by

$$\varrho_{\text{cm}} = 2c \, \frac{\displaystyle\int_{u_{\min}}^{u_{\max}} |u| \, v^2(u) \, \mathrm{d}u}{\displaystyle\int_{u_{\min}}^{u_{\max}} v^2(u) \, \mathrm{d}u} \, . \tag{7.62}$$

7.3 Cassinian Ovaloids

7.3.1 Symmetric Shapes

A one-parameter parameterization which covers the sphere, constricted, scission, and separated shapes is the family of Cassinian ovaloids [Adeev 71], [Pashkevich 71], [Stavinsky 68]

$$\varrho_s^2 = \sqrt{a^4 + 4c^2 z^2} - (c^2 + z^2) \, . \tag{7.63}$$

In using

$$u = c/a \, ,$$

a is eliminated by volume conservation,

$$4 \left(\frac{R_0}{a} \right)^3 = \begin{cases} \sqrt{1 + u^2}(1 - 2u^2) + \dfrac{3}{2u} \operatorname{arsinh}(2u\sqrt{1 + u^2}) & , \ u \leq 1 \\[2ex] \dfrac{3}{2u} \operatorname{arsinh}(ut) - t \\ \text{with } t = (2u^2 - 1)\sqrt{1 + u^2} - (2u^2 + 1)\sqrt{u^2 - 1} & , \ u \geq 1 \, . \end{cases} \tag{7.64}$$

Half length and neck radius for connected shapes,

$$\left. \begin{aligned} z_0 &= \sqrt{a^2 + c^2} \\[1ex] \varrho_n &= \sqrt{a^2 - c^2} \end{aligned} \right\} \quad \text{for } u \geq 1 \, . \tag{7.65}$$

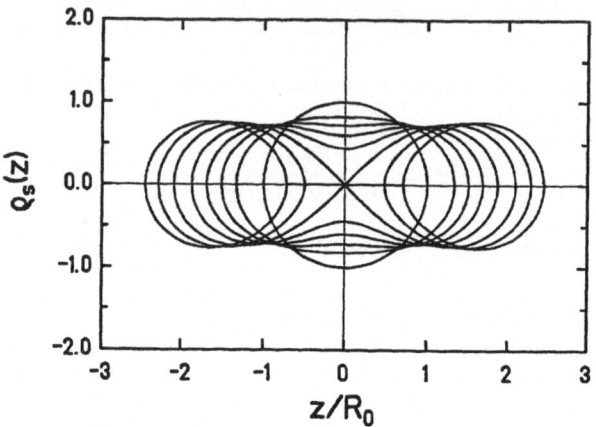

Figure 7.4: Cassinian ovaloids. The parameter u^4 varies from 0 (one sphere) to 1.4 with a step of 0.2. Two tangent fragments correspond to $u = 1$.

Special cases:

$$
\begin{array}{ll}
u = 0 & \text{, sphere} \\
0 < u < 2^{-1/2} & \text{, similar to spheroid} \\
2^{-1/2} < u < 1 & \text{, necked in shapes} \\
u = 1 & \text{, scission (lemniscate)} \\
u > 1 & \text{, two disconnected shapes .}
\end{array}
$$

Figure 7.4 shows the shapes contained in the parameterization of symmetric Cassinian ovaloids. Another notation which is used in [Stavinsky 68]:

$$
\begin{aligned}
s &= c^2 \\
R &= a \\
\varepsilon &= (c/R_0)^2 .
\end{aligned}
\tag{7.66}
$$

7.3.2 Distorted Shapes

Cassinian ovaloids can be distorted in two ways, either [Adeev 71]

$$
\varrho_s^2 = \sqrt{a^4 + 4c^2 z^2} - (c^2 - \varepsilon^2 + z^2) ,
\tag{7.67}
$$

or [Pashkevich 71] define the Cassinian radial and angular variables, cf. [Moon 61],

$$
a(\varrho, z) = \left[(z^2 + \varrho^2)^2 + 2c^2(\varrho^2 - z^2) + c^4 \right]^{1/4}
$$

$$
\mu(\varrho, z) = \frac{\text{sign}(z)}{\sqrt{2}} \left[1 + \frac{z^2 - \varrho^2 - c^2}{a^2(\varrho, z)} \right]^{1/2}
\tag{7.68}
$$

and perturb the radial coordinate harmonically ,

84

$$a(\mu) = a_0 \left(1 + \sum_{n=1}^{N} a_n \mathrm{P}_n(\mu)\right). \tag{7.69}$$

Appropriate formulae for the surface and Coulomb energies can be found in [Strutinsky 63], [Pashkevich 71].

7.3.3 Lemniscatoids

Descendants of Cassinian ovaloids are the lemniscatoids [Royer 82], [Royer 84] obtained by inverting an ellipsoid. In cylindrical coordinates

$$A^2 \varrho_s^2 + C^2 z^2 = (\varrho_s^2 + z^2)^2 \tag{7.70}$$

and in spherical coordinates

$$R^2(\theta) = A^2 \sin^2 \theta + C^2 \cos^2 \theta. \tag{7.71}$$

If the single variable $s = A/C$ is introduced then $s = 1$ corresponds to a single sphere and $s = 0$ is the configuration of two tangent spheres, see Fig. 7.5. Volume conservation demands

$$\begin{aligned} V &= \frac{4}{3}\pi R_0^3 \\ &= \frac{1}{12}\pi C^3 \left[4 + 6s^2 + \frac{3s^4}{\sqrt{1-s^2}} \operatorname{arsinh} 2 \frac{\sqrt{1-s^2}}{s^2}\right]. \end{aligned} \tag{7.72}$$

The distance between the centers of mass of the two portions is given by

$$R_{\mathrm{cm}} = \pi C^4 \frac{1 + s^2 + s^4}{3V} \tag{7.73}$$

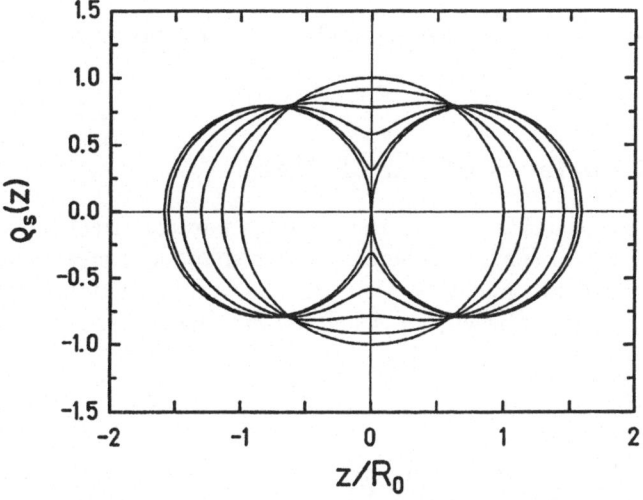

Figure 7.5: Lemniscatoids. The parameter s varies from 0 (two tangent spheres) to 1 (one sphere) with a step of 0.2.

and the relative surface and curvature energies read

$$B_{\text{surf}} = \frac{C^2}{2R_0^2} \left[1 + \frac{s^4}{\sqrt{1-s^4}} \operatorname{arsinh} \frac{\sqrt{1-s^4}}{s^2} \right] \tag{7.74}$$

$$\begin{aligned} B_{\text{curv}} = \ & \frac{C}{R_0} \left[1 + \frac{s^4}{2(1+s^2)\sqrt{1-s^2}} \operatorname{arsinh} \frac{\sqrt{1-s^2}}{s} \right. \\ & \left. - \frac{s}{2(1+s^2)\sqrt{1-s^2}} \arctan \frac{\sqrt{1-s^2}}{s} \right]. \end{aligned} \tag{7.75}$$

The Coulomb energy has to be calculated numerically. In the natural units of Chap. 1 the perpendicular moment of inertia and quadrupole moment read

$$\begin{aligned} \mathcal{J}_\perp = \ & \frac{C^5}{R_0^5} \frac{s^2}{512(1-s^2)} \\ & \times \left[\frac{112}{s^2} + 8 + 30s^2 - 135s^4 + \frac{120s^4 - 135s^6}{\sqrt{1-s^2}} \operatorname{arsinh} \frac{\sqrt{1-s^2}}{s} \right] \end{aligned} \tag{7.76}$$

$$\begin{aligned} Q = \ & \frac{C^5}{R_0^5} \frac{\pi s^2}{96(1-s^2)} \\ & \times \left[\frac{16}{s^2} - 8 - 14s^2 + 15s^4 - \frac{24s^4 - 15s^6}{\sqrt{1-s^2}} \operatorname{arsinh} \frac{\sqrt{1-s^2}}{s} \right]. \end{aligned} \tag{7.77}$$

Lemniscatoids can also be distorted harmonically to yield more shape degrees of freedom, for instance by employing

$$R^2(\theta) = R_0^2 \lambda^{-1} \left(1 + \sum_{n=1}^{N} c_n P_n(\cos\theta) \right), \tag{7.78}$$

where λ stands for volume conservation and $n = 2$ corresponds to pure lemniscatoids.

7.4 Matched Surfaces of Revolution

Except for the lemniscatoids none of the previous parameterizations contains the configuration of two tangent spheres. As this shape is desirable if dealing with the scattering of two nuclei or with certain compact fission trajectories, one has to sacrifice the analyticity of the shape function.

7.4.1 Matched Spheres

7.4.1.1 Symmetric Shapes

In cylindrical coordinates, two overlapping or separated spheres are given by [Grammaticos 73], [Hasse 77]

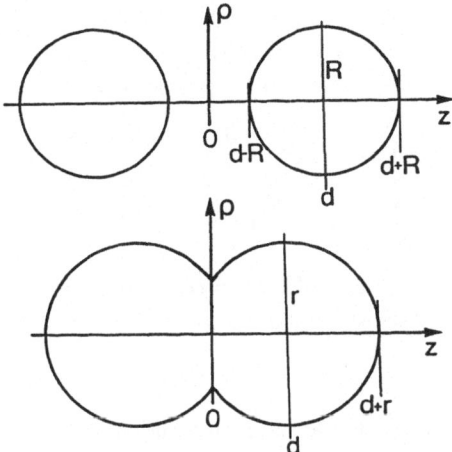

Figure 7.6: The two sphere parameterization for overlapping and separated fragments (from [Hasse 77]).

$$\varrho_s^2 = r^2 - (\mid z \mid + d)^2 .\tag{7.79}$$

The meaning of the parameters r, d can be seen from Fig. 7.6 and

$$R = 2^{-1/3} R_0 .\tag{7.80}$$

Use

$$\varepsilon = 1 - d/r\tag{7.81}$$

as the only parameter and eliminate d by

$$r = \begin{cases} d \left[\cosh\left(\frac{1}{3}\operatorname{arcosh}(1 + 4R_0^3/d^3)\right) - \frac{1}{2}\right] & , \varepsilon \geq 0 \\ R & , \varepsilon \leq 0 \end{cases}\tag{7.82}$$

from volume conservation

$$d^3 - 3dr^2 + 2R_0^3 - 2r^3 = 0 \quad \text{for } \varepsilon \geq 0 .\tag{7.83}$$

The neck radius and half length are:

$$\begin{aligned} \varrho_n &= \sqrt{r^2 - d^2} \\ z_0 &= d + r . \end{aligned}\tag{7.84}$$

Special cases

$$\begin{array}{ll} \varepsilon = 1 & , \text{ single sphere} \\ 0 < \varepsilon < 1 & , \text{ two overlapping spheres} \\ \varepsilon = 0 & , \text{ two tangent spheres} \\ \varepsilon < 0 & , \text{ two separated spheres} . \end{array}$$

The system of two overlapping or separated spheres can also be described in toroidal coordinates [Brosa 80], where the hydrodynamic mass parameters of Sect. 1.8 are calculated exactly.

In Werner-Wheeler approximation (cited in [Nix 68], [Nix 69]) the mass parameter and two-body viscosity coefficients of Sect. 1.8 with respect to the deformation parameter d/R_0 are given by [Hasse 77]

$$B_d = \left(\frac{r}{R_0}\right)^3 \left(1 - \frac{21}{8}\epsilon + \frac{3}{4}\epsilon^2 + \frac{5}{32}\epsilon^3 - \frac{9}{8}\epsilon^2 \log\frac{\epsilon}{2}\right)$$

$$Z_d = \frac{r}{3R_0}\left(\frac{1}{\epsilon} + 3 - \frac{15}{4}\epsilon + \epsilon^2\right). \tag{7.85}$$

The Coulomb interaction energy of two overlapping or separated spheres are given by lengthy expressions in the appendix of [Devries 75].

7.4.1.2 Asymmetric Shapes

In dealing with asymmetric overlapping or tangent spheres, it is more convenient to employ plane toroidal coordinates (η, θ) defined by (cf. [Moon 61])

$$z = \frac{w \sin\theta}{\cosh\eta - \cos\theta}$$

$$\varrho = \frac{w \sinh\eta}{\cosh\eta - \cos\theta}. \tag{7.86}$$

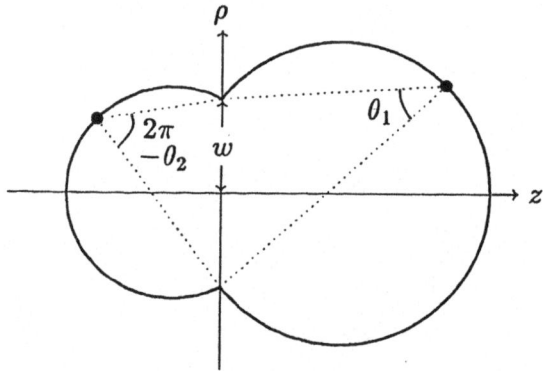

Figure 7.7: Angles and orifice in the parameterization of two overlapping spheres.

According to Fig. 7.7 w is the radius of the orifice and points on the left and right spheres are described by constant angles θ_1 or θ_2, hence $w = w(\theta_1)$ or $w(\theta_2)$. Introducing the asymmetry parameter

$$a = \frac{v_1 - v_2}{V}, \tag{7.87}$$

where v_1, v_2, are the volumes of the left and right sphere, respectively, and $V = 4\pi R_0^3/3$, volume conservation yields

$$a + 1 = f(\theta_1)$$
$$a - 1 = f(\theta_2)$$
$$f(\theta_i) = \frac{1}{4}w^3\left(3 + \cot\frac{\theta_i}{2}\right)\cot\frac{\theta_i}{2} \quad , \; i = 1,2 \, . \qquad (7.88)$$

Hence (w, a) is equivalent to (θ_1, θ_2). Special cases are

$$w = R_0, \quad a = 0; \; \theta_1 = \frac{\pi}{2}, \quad \theta_2 = \frac{3\pi}{2} \qquad \text{, single sphere}$$
$$a = 0, \quad \theta_2 = 2\pi - \theta_1 \qquad \text{, symmetric shapes}$$
$$\theta_1 \approx w\left(\frac{1+a}{2}\right)^{-1/3}, \quad \theta_2 \approx 2\pi - w\left(\frac{1-a}{2}\right)^{-1/3} \qquad \text{, slightly overlapping}$$
$$\text{spheres} \, .$$

7.4.2 Dumbbells

In adding a cylindrical neck between two portions of a sphere, one obtains the dumbbell parameterization [Hasse 77], [Swiatecki 80] . Using the notation of Fig. 7.8 the surface is given by

$$\varrho_s^2 = \begin{cases} u^2 - (z + \sigma_0)^2 & , \; -\sigma_0 - u \leq z \leq -\sigma_0 + \sigma_2 \\[2mm] u^2 - \sigma_2^2 & , \; -\sigma_0 + \sigma_2 \leq z \leq +\sigma_0 - \sigma_2 \\[2mm] u^2 - (z - \sigma_0)^2 & , \; +\sigma_0 - \sigma_2 \leq z \leq +\sigma_0 + u \, . \end{cases} \qquad (7.89)$$

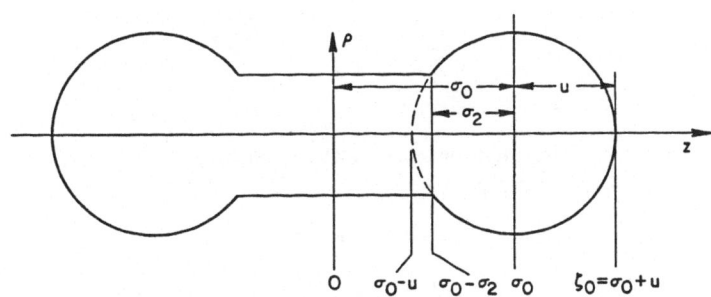

Figure 7.8: The dumbbell-parameterization (from [Hasse 71]).

The radius u is eliminated by volume conservation

$$u = \begin{cases} \left(-q + \sqrt{q^2 + p^3}\right)^{1/3} \\ \quad + \left(-q - \sqrt{q^2 + p^3}\right)^{1/3} - \frac{1}{2}\sigma_0 \qquad , \; \Delta \geq 0 \\[3mm] \sigma_0\left[\cos\left(\frac{1}{3}\arccos\left(-\frac{8q}{\sigma_0^3}\right)\right) - \frac{1}{2}\right] \quad , \; \Delta \leq 0 \end{cases} \qquad (7.90)$$

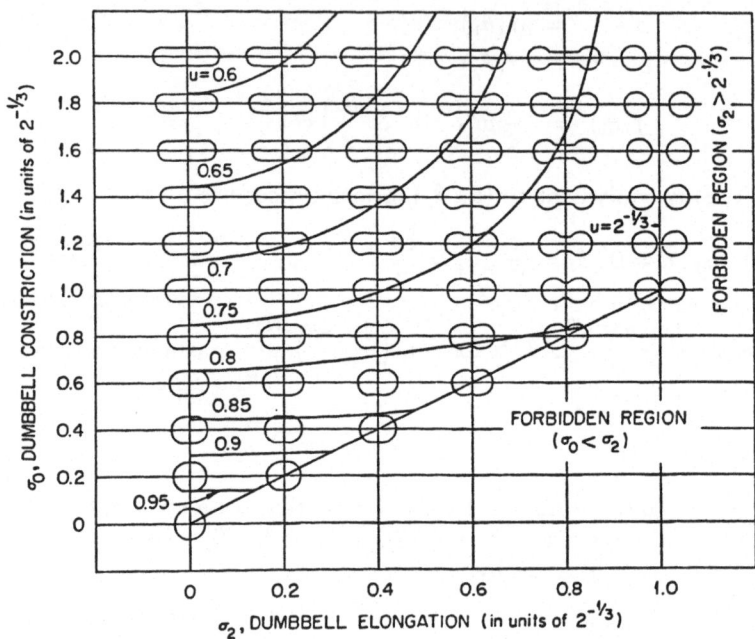

Figure 7.9: Shapes contained in the dumbbell-parameterization (from [Hasse 71]).

where

$$p = -\left(\frac{\sigma_0}{2}\right)^2$$

$$q = \left(\frac{\sigma_0}{2}\right)^3 - \frac{3}{2}\frac{\sigma_0}{2}\sigma_2^2 + \frac{1}{2}\sigma_2^3 - \frac{1}{2}$$

$$\Delta = q^2 + p^3 . \tag{7.91}$$

Dumbbell shapes in this parameterization are shown in Fig. 7.9. The relative surface and curvature energies are

$$B_{\text{surf}} = u(u + \sigma_2) + \varrho_n(\sigma_0 - \sigma_2) \tag{7.92}$$

$$B_{\text{curv}} = u + \frac{1}{2}\left(\sigma_0 + \sigma_2 - \varrho_n \arctan\frac{\sigma_2}{\varrho_n}\right) , \tag{7.93}$$

where

$$\varrho_n = \sqrt{u^2 - \sigma_2^2} \tag{7.94}$$

is the cylindrical neck radius. The Coulomb energy can be evaluated with the help of [Gaudin 74], [Martinot 77]. For instance, the relative Coulomb self energy (i.e. in units of the Coulomb energy of the volume equivalent sphere) of a cylinder of radius a, length 2ℓ and $\beta = \ell/a$ is

$$B_{\text{Coul}} = \frac{5}{2\pi}\left(\frac{2}{3}\right)^{2/3}\frac{D(\beta)}{\beta^{5/3}} , \tag{7.95}$$

where

90

$$D(\beta) = \frac{\pi}{2}\beta \log \beta + \pi\beta \left(\log 2 - \frac{3}{8}\right) + \frac{32}{45}$$

$$- \frac{\pi}{32} \sum_{n=0}^{\infty} \frac{\left(\frac{1}{2}\right)_n \left(\frac{5}{2}\right)_n}{(3)_n (4)_n} \frac{(-)^n}{(n+1)\beta^{2n+1}} \tag{7.96}$$

and

$$(x)_n = \frac{\Gamma(x+n)}{\Gamma(x)} \tag{7.97}$$

is the Pochhammer symbol. It is of interest to note that the relative Coulomb energy of a hemisphere [Carlson 63] (erroneous in [Krappe 81]),

$$B_{\text{Coul}} = \frac{32 - 3\pi}{6\pi\,2^{1/3}} = 0.9506, \tag{7.98}$$

is within 5% of that of the full sphere.

7.4.3 Matched Quadratic Surfaces

The most versatile parameterization is the one of Nix [Nix 68], [Nix 69], [Nix 72] who used three smoothly joined portions of quadratic surfaces of revolution, i.e. spheres, spheroids, hyperboloids, cylinders and cones

$$\varrho_s^2 = \begin{cases} a_1^2 - (a_1^2/c_1^2)(z-l_1)^2 & , \quad l_1 - c_1 \leq z \leq z_1 \\ a_2^2 - (a_2^2/c_2^2)(z-l_2)^2 & , \quad z_2 \leq z \leq l_2 + c_2 \\ a_3^2 - (a_3^2/c_3^2)(z-l_3)^2 & , \quad z_1 \leq z \leq z_2 \ . \end{cases} \tag{7.99}$$

For the geometry , cf. Fig. 7.10. From the 11 original parameters a_i, c_i, l_i (i=1,2,3), z_1, z_2, four are eliminated by matching the surfaces and their derivatives at the points z_1, z_2, and one more parameter is eliminated by volume conservation. In principle, another one could be eliminated by center-of-mass conservation. As this turns out to be too complicated, it is usually retained

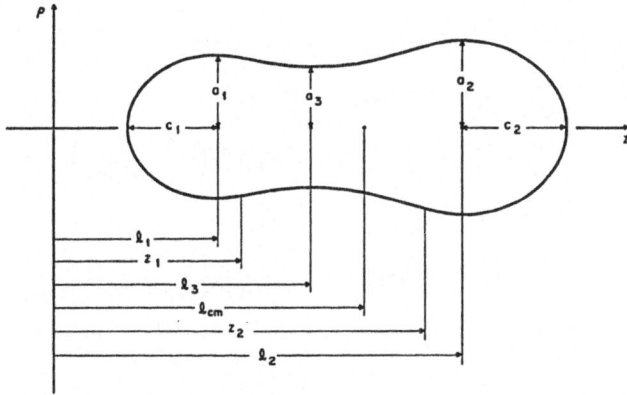

Figure 7.10: The parameters of the matched quadratic surfaces (from [Nix 68]).

and eliminated after having performed the calculations. One is, hence, left with three symmetric parameters,

$$\sigma_1 = (l_2 - l_1)/u \ ,$$

$$\sigma_2 = a_3^2/c_3^2 \ ,$$

$$\sigma_3 = \frac{1}{2}\left[\left(a_1^2/c_1^2\right) + \left(a_2^2/c_2^2\right)\right] \tag{7.100}$$

and three asymmetrical ones

$$\alpha_1 = \frac{1}{2}(l_1 + l_2)/u \ ,$$

$$\alpha_2 = (a_1^2 - a_2^2)/u^2 \ ,$$

$$\alpha_3 = (a_1^2/c_1^2) - (a_2^2/c_2^2) \quad , \tag{7.101}$$

where

$$u = \sqrt{\frac{1}{2}(a_1^2 + a_2^2)} \ . \tag{7.102}$$

Selected symmetric middle portions, $z_1 \leq z \leq z_2$ are shown in Fig. 7.11.

Inversion of the eleven original parameters in terms of the six new ones proceeds as follows. Seven are given by

$$l_1 = \frac{1}{2}(-\sigma_1 + 2\alpha_1)u$$

$$l_2 = \frac{1}{2}(\sigma_1 + 2\alpha_1)u$$

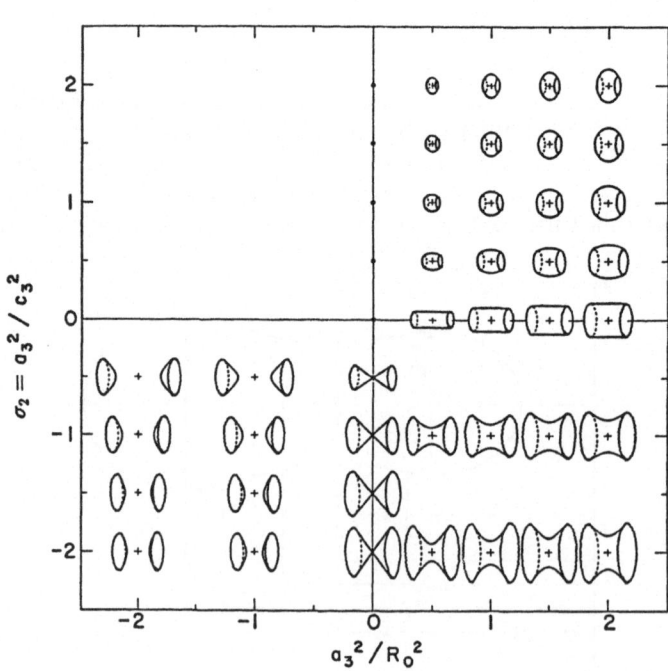

Figure 7.11: Selected middle portions of the parameterization of matched quadratic surfaces (from [Nix 68]).

92

$$a_1^2 = (1+\frac{1}{2}\alpha_2)u^2$$

$$a_2^2 = (1-\frac{1}{2}\alpha_2)u^2$$

$$a_1^2/c_1^2 = \sigma_3+\frac{1}{2}\alpha_3$$

$$a_2^2/c_2^2 = \sigma_3-\frac{1}{2}\alpha_3$$

$$a_3^2/c_3^2 = \sigma_2 \ . \tag{7.103}$$

The matching conditions yield

$$a_i^2 - (a_i^2/c_i^2)(z_i - l_i)^2 = a_3^2 - (a_3^2/c_3^2)(z_i - l_3)^2 \ , \quad i = 1,2, \tag{7.104}$$

$$z_i = \frac{(a_i^2/c_i^2)l_i - (a_3^2/c_3^2)l_3}{(a_i^2/c_i^2) - (a_3^2/c_3^2)} \ , \quad i = 1,2 \tag{7.105}$$

which still contain l_3. Hence

$$l_3 = 0 \quad \text{for} \ \alpha_3 = 0 \ \text{(symmetric shapes)} \tag{7.106}$$

or

$$\sigma_2\alpha_3 l_3 = \text{sign}(B)\sqrt{B^2 - AC} - B \ , \tag{7.107}$$

where

$$
\begin{aligned}
A &= (a_1^2/c_1^2) - (a_2^2/c_2^2) \\
B &= (a_1^2/c_1^2)\left[(a_2^2/c_2^2) - (a_3^2/c_3^2)\right] l_1 \\
&\quad - (a_2^2/c_2^2)\left[(a_1^2/c_1^2) - (a_3^2/c_3^2)\right] l_2 \\
C &= (a_3^2/c_3^2)\left\{(a_2^2/c_2^2)\left[(a_1^2/c_1^2) - (a_3^2/c_3^2)\right] l_2^2 \right. \\
&\quad \left. - (a_1^2/c_1^2)\left[(a_2^2/c_2^2) - (a_3^2/c_3^2)\right] l_1^2\right\} \\
&\quad + \left[(a_1^2/c_1^2) - (a_3^2/c_3^2)\right]\left[(a_2^2/c_2^2) - (a_3^2/c_3^2)\right](a_2^2 - a_1^2) \ . \tag{7.108}
\end{aligned}
$$

The quantities z_i can now be computed from (7.105). To account for volume conservation, u is eliminated by

$$\left(\frac{R_0}{u}\right)^3 = \frac{3}{4\pi}(\tilde{V}_1 + \tilde{V}_2 + \tilde{V}_3) \ , \tag{7.109}$$

where $\tilde{V}_i$ are the volumes of the respective portions scaled in such a way that $u = 1$,

$$\tilde{V}_i = \frac{1}{2}\tilde{a}_i^2\tilde{c}_i + (-1)^i\left[\frac{3}{4}\tilde{a}_i^2(\tilde{l}_i - \tilde{z}_i) - \frac{1}{4}(\tilde{a}_i^2/\tilde{c}_i^2)(\tilde{l}_i - \tilde{z}_i)^3\right] \ , \quad i = 1,2$$

$$\tilde{V}_3 = \frac{3}{4}\tilde{a}_3^2(\tilde{z}_2 - \tilde{z}_1) - \frac{1}{4}(\tilde{a}_3^2/\tilde{c}_3^2)\left[(\tilde{l}_3 - \tilde{z}_1)^3 + (\tilde{z}_2 - \tilde{l}_3)^3\right] - \tilde{a}_3^2\tilde{c}_3\,\Theta(-\tilde{a}_3^2) \ ,$$

$$\tag{7.110}$$

where $\Theta(x)$ denotes the step function. The tilde also denotes scaled quantities, e.g.

$$\tilde{l}_1 = \frac{1}{2}(-\sigma_1 + 2\alpha_1)$$

$$\tilde{c}_1 = \left[\tilde{a}_1^2/(\tilde{a}_1^2/\tilde{c}_1^2)\right]^{\frac{1}{2}}$$

$$= \left[(1 + \frac{1}{2}\alpha_2)/(\sigma_3 + \frac{1}{2}\alpha_3)\right]^{\frac{1}{2}}. \tag{7.111}$$

Location of the center-of-mass:

$$
\begin{aligned}
l_{cm}R_0^3 = {} & \frac{3}{8}a_1^2\left[z_1^2 - (l_1 - c_1)^2\right] - (a_1^2/c_1^2)\left\{\frac{3}{16}\left[z_1^4 - (l_1 - c_1)^4\right]\right. \\
& \left. - \frac{1}{2}l_1\left[z_1^3 - (l_1 - c_1)^3\right] + \frac{3}{8}l_1^2\left[z_1^2 - (l_1 - c_1)^2\right]\right\} \\
& + \frac{3}{8}a_2^2\left[(l_2 + c_2)^2 - z_2^2\right] - (a_2^2/c_2^2)\left\{\frac{3}{16}\left[(l_2 + c_2)^4 - z_2^4\right]\right. \\
& \left. - \frac{1}{2}l_2\left[(l_2 + c_2)^3 - z_2^3\right] + \frac{3}{8}l_2^2\left[(l_2 + c_2)^2 - z_2^2\right]\right\} \\
& + \frac{3}{8}a_3^2(z_2^2 - z_1^2) - (a_3^2/c_3^2)\left[\frac{3}{16}(z_2^4 - z_1^4)\right. \\
& \left. - \frac{1}{2}l_3(z_2^3 - z_1^3) + \frac{3}{8}l_3^2(z_2^2 - z_1^2)\right]. \tag{7.112}
\end{aligned}
$$

7.4.4 Other Matched Surfaces

A subset of the **Matched Quadratic Surfaces** parameterization described in the previous section, consisting of two spheres smoothly connected by a quadratic surface, has been employed by Błocki and Swiatecki in [Błocki 82]. This 865 page encyclopedic compilation of potential energy surfaces for heavy-ion reactions employs the three primary parameters

$$\text{distance} \qquad \varrho = \frac{r}{R_1 + R_2}$$

$$\text{"deck parameter"} \qquad \lambda = \frac{l_1 + l_2}{R_1 + R_2}$$

$$\text{asymmetry} \qquad \Delta = \frac{R_1 - R_2}{R_1 + R_2}, \tag{7.113}$$

where R_1 and R_2 are the radii of the spheres, r is the center to center distance between them and l_i is the thickness of the lens-shaped piece of the sphere that lies within the matching quadratic surface, cf. Fig. 7.12.

For divided shapes the *deck parameter* λ (*deck* is a combination of the words *deformation* and *neck*) is a measure of the bulging (*polarization* or *nose-forma-tion*) of the approaching fragments. After contact, λ goes over into a measure of the degree of opening of the neck or window through which the fragments communicate. As λ increases further the neck concavity disappears and the shapes become convex everywhere. Here $\lambda = 1 - 1/\varrho$ corresponds to the scission line with $\varrho = 1$, $\lambda = 0$ being the configuration of two tangent spheres; $\lambda =$

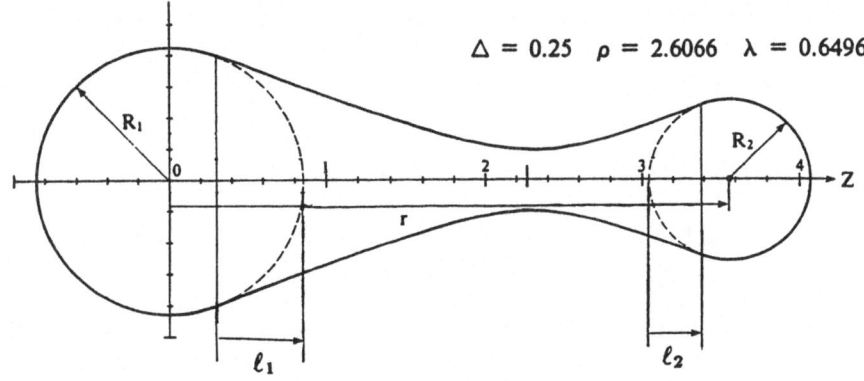

$$\Delta = 0.25 \quad \rho = 2.6066 \quad \lambda = 0.6496$$

Figure 7.12: The parameters of the Błocki- Swiatecki- parameterization (from [Błocki 82]).

$2 - \Delta - \Delta/\varrho$ corresponds to single egg-like ovaloids and $\lambda = 2 + \Delta + \Delta/\varrho$ are single spheroids. Some examples of these shapes are shown in Fig. 7.13. An earlier study along similar lines [Swiatecki 81] employed a parameterization consisting of two spheres connected by a cone having various opening angles.

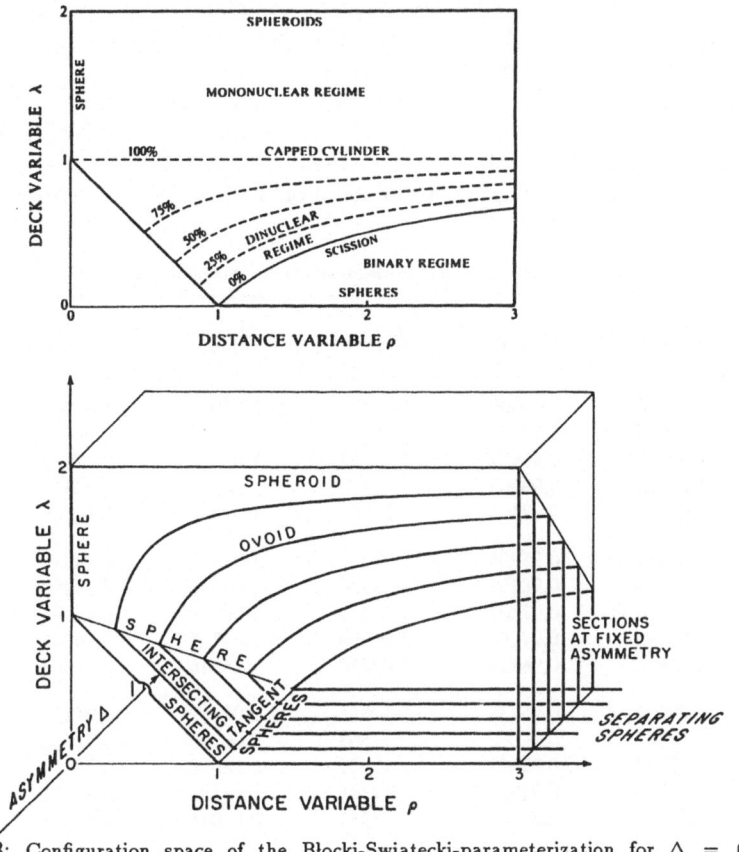

Figure 7.13: Configuration space of the Błocki-Swiatecki-parameterization for $\Delta = 0$ (from [Błocki 82]).

Other parameterizations that appear in the literature are: Matched cones or matched cylinders [Cohen 62], overlapping spheroids [Nix 64], [Nix 65], and two portions of a spheroid connected by a surface of sixth power in z [Guet 80].

7.5 y-Family

In the caption to the second figure of their classic work on fission [Hill 53], Hill and Wheeler propose a one-dimensional family of shapes for use in the discussion of fission barriers. They observed that in the liquid drop model fission barrier saddle point shapes for values of the fissility parameter x, cf. Eq. (1.75), decreasing from one toward zero form a sequence that could also be used to describe the path, for a particular nucleus, from the spherical ground state up over the fission barrier and down toward scission. They stress that when the saddle point shapes are used in this way they need not be considered to have anything to do with the question of equilibrium forms. To help facilitate this separation they proposed to associate each equilibrium shape with a parameter y, equal to one minus the value of the fissility parameter x that it corresponds to

$$y = 1 - x . \tag{7.114}$$

Thus y can be used to describe the shape of a nucleus which can have any value of the quantity $x \propto (Z^2/A)$. Of course, the true LDM saddle point shape for a nucleus with fissility x is a y-family shape.

The properties of this family of shapes are also discussed in [Swiatecki 58], where the symbol t is used instead of y. In addition, the y-family shapes are used extensively in [Bolsterli 72] where fission barriers are calculated for heavy and superheavy nuclei using a combination of the LDM and single-particle shell effects.

Table 7.1: Numerical values of the relative energies (from [Myers 74]).

y	B_{surf}	B_{Coul}	B_{curv}	B_{red}	$B_v = \sqrt{B_{\text{sr1}}}$	$B_w = B_{\text{sr2}}$
0.00	1.00000	1.00000	1.00000	1.00000	1.00000	1.00000
0.02	1.00086	0.99957	1.00087	1.00085	0.99957	1.00000
0.04	1.00338	0.99827	1.00352	1.00341	0.99826	0.99995
0.06	1.00750	0.99609	1.00799	1.00746	0.99605	0.99977
0.08	1.01319	0.99303	1.01433	1.01267	0.99285	0.99927
0.10	1.02044	0.98905	1.02265	1.01857	0.98855	0.99819
0.12	1.02927	0.98409	1.03306	1.02446	0.98298	0.99619
0.14	1.03974	0.97807	1.04576	1.02944	0.97591	0.99278
0.16	1.05195	0.97088	1.06099	1.03232	0.96706	0.98736
0.18	1.06604	0.96239	1.07910	1.03146	0.95604	0.97908
0.20	1.08224	0.95238	1.10056	1.02469	0.94238	0.96685
0.22	1.10085	0.94060	1.12603	1.00906	0.92546	0.94915
0.24	1.12229	0.92667	1.15651	0.98048	0.90450	0.92390
0.26	1.14717	0.91008	1.19348	0.93335	0.87854	0.88812
0.28	1.17623	0.89017	1.23915	0.86045	0.84669	0.83771

96

y	B_{surf}	B_{Coul}	B_{curv}	B_{red}	$B_v=\sqrt{B_{\mathrm{sr1}}}$	$B_w=B_{\mathrm{sr2}}$
0.30	1.20963	0.86664	1.29590	0.75626	0.80959	0.76880
0.32	1.24296	0.84250	1.35951	0.63714	0.77505	0.58918
0.34	1.26532	0.82584	1.41013	0.55370	0.75660	0.62984
0.36	1.27619	0.81749	1.44103	0.51320	0.75132	0.59866
0.38	1.28126	0.81347	1.46026	0.49390	0.75123	0.58281
0.40	1.28362	0.81155	1.47339	0.48414	0.75297	0.57442
0.42	1.28458	0.81073	1.48308	0.47907	0.75532	0.56995
0.44	1.28477	0.81057	1.49068	0.47657	0.75783	0.56774
0.46	1.28450	0.81081	1.49692	0.47559	0.76032	0.56694
0.48	1.28394	0.81134	1.50226	0.47557	0.76272	0.56706
0.50	1.28320	0.81206	1.50694	0.47620	0.76497	0.56783
0.52	1.28235	0.81294	1.51114	0.47730	0.76709	0.56906
0.54	1.28141	0.81394	1.51502	0.47875	0.76906	0.57065
0.56	1.28042	0.81503	1.51864	0.48048	0.77090	0.57252
0.58	1.27941	0.81622	1.52208	0.48245	0.77260	0.57463
0.60	1.27837	0.81748	1.52539	0.48463	0.77419	0.57696
0.62	1.27732	0.81882	1.52861	0.48703	0.77566	0.57948
0.64	1.27627	0.82024	1.53177	0.48964	0.77703	0.58220
0.66	1.27522	0.82174	1.53490	0.49246	0.77830	0.58513
0.68	1.27418	0.82333	1.53803	0.49553	0.77949	0.58827
0.70	1.27314	0.82501	1.54117	0.49886	0.78059	0.59166
0.72	1.27210	0.82679	1.54473	0.50248	0.78163	0.59531
0.74	1.27108	0.82869	1.54762	0.50645	0.78259	0.59929
0.76	1.27006	0.83072	1.55096	0.51082	0.78350	0.60364
0.78	1.26906	0.83291	1.55440	0.51568	0.78435	0.60843
0.80	1.26806	0.83528	1.55798	0.52111	0.78516	0.61377
0.82	1.26707	0.83788	1.56173	0.52728	0.78594	0.61979
0.84	1.26610	0.84074	1.56567	0.53436	0.78669	0.62669
0.86	1.26514	0.84396	1.56980	0.54265	0.78742	0.63472
0.88	1.26418	0.84763	1.57413	0.55256	0.78816	0.64430
0.90	1.26325	0.85190	1.57860	0.56475	0.78892	0.65606
0.92	1.26233	0.85699	1.58301	0.58024	0.78973	0.67097
0.94	1.26147	0.86321	1.58688	0.60062	0.79063	0.69056
0.96[b]						
0.98[b]						
1.00[c]	1.25992	0.89244	1.58740	0.72236	0.79370	0.80816

[a] Calculated with the shape parameterization of matched quadratic surfaces.

[b] The shape parameterization fails in this region.

[c] Calculated for the configuration of tangent spheres.

The LDM deformation energy $B_{\text{Def}}(x,y)$ (in units of the surface energy) can be written as

$$B_{\text{Def}}(x,y) = (B_{\text{surf}}(y) - 1) + 2x(B_{\text{Coul}}(y) - 1). \qquad (7.115)$$

The quantities B_{surf} and B_{Coul} are the relative surface and Coulomb energies for the LDM saddle point shapes which are tabulated in various places, for example see Table 7.1.

In Fig. 7.14 the quantity $B_{\text{Def}}(x,y)$ is plotted versus y for a number of discrete values of x. It should be noted that these curves each have three extrema, one at the $y = 0$ spherical shape, one at the top of the fission barrier where $y = x$, and another at $y = 0.437$. The stationary point at $y = 0.437$ is associated with the fact that the derivatives of both B_{surf} and B_{Coul} with respect to x (or y) change sign at this point. (Of course, if either B_{surf} or B_{Coul} is stationary the other must be as well.)

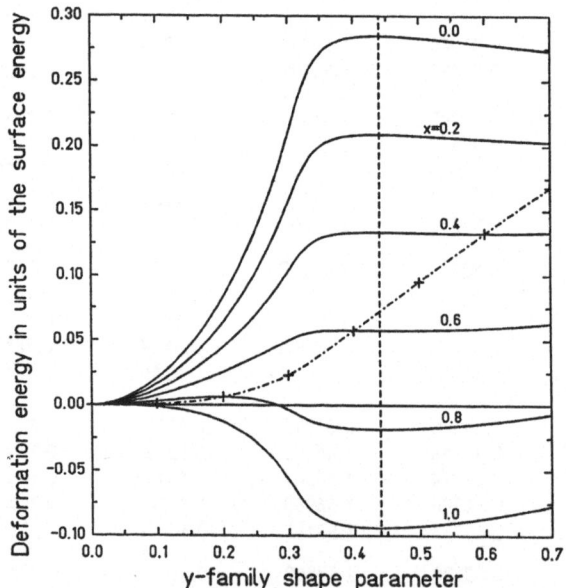

Figure 7.14: The quantity $B_{\text{Def}}(x,y)$, which is the deformation energy in units of the surface energy of the sphere, is plotted versus y for various values of x. The vertical dashed line locates the extremum at $y = 0.437$ and the dot-dashed line is the locus of the extrema corresponding to the fission saddle point shapes. For each x it lies at $y = 1 - x$.

The somewhat confusing fact that the true LDM saddle point lies at a minimum (not a maximum) on the curves for $x < 0.563$ means that for these x-values the y-family shapes do not *pass over* the saddle point but instead traverse it from *side to side*. For $x = 0.563$ the saddle shape is at an inflection point and for this particular x value the y-family traverses the saddle point along an equipotential.

Since tabulated values of the relative surface energy, Coulomb energy, curvature energy, redistribution energy, etc. are available for y-family shapes, see Table 7.1, it is natural to think of using this one-dimensional family for estimating the barrier energy when various small corrections to the LDM are also included. This procedure was used in [Myers 77] where special care was required to identify the saddle point when it corresponded to a minimum (or a point of inflection) in $B_{\mathrm{Def}}(y)$.

Chapter 8

Saddle Point Properties

8.1 Liquid Drop Barriers

For reasonable shapes $\{\alpha\}$ described by a set of dimensionless deformation parameters the liquid drop deformation energy B_{Def} exhibits a barrier for each value of the fissility x, cf. Eq. (1.75), for instance the configuration of two tangent spheres for $x = 0$ and a single sphere for $x = 1$. The deformation energy at the saddle deformation $\{\hat{\alpha}\}$ is called $B_{\text{Bar}}(x, \hat{\alpha})$ and corresponds, if multiplied by E_S^0, to the LDM fission barrier height. The symbol ξ is sometimes also used for B_{Bar}, for instance in [Wilets 64].

Liquid Drop barrier shapes and energies can also be found in a deformation parameter-free and therefore exact way with the variational method of Strutinsky [Strutinsky 62], [Strutinsky 63], cf. Sect. 7.1.1.2, by solving the integro-differential equation (7.18). The Lagrange parameter λ_1 here vanishes because the saddle point is a stationary point and, hence, the deformation constraining function $f(\zeta, \varrho)$ becomes irrelevant.

For large x close to unity, i.e. $y = 1 - x \ll 1$ saddle point properties can be studied analytically. Using the spheroidal expansion of Chap. 6 the saddle point (here a maximum in one dimension) eccentricity reads [Hasse 71]

$$\hat{e}^2 = 7y\,(1 - y \cdots) \tag{8.1}$$

and the relative barrier energy is given by [Hill 53], [Businaro 55a]

$$B_{\text{Bar}} = \frac{98}{135}\, y^3 \left(1 - \frac{64}{3}\, y \cdots\right) \tag{8.2}$$

In the α_n-parameterization the lowest parameters and the relative barrier are given by [Swiatecki 58]

$$
\begin{aligned}
\hat{\alpha}_2 &= \frac{7}{3}\, y - \frac{938}{765}\, y^2 + 9.499768\, y^3 - 8.050944\, y^4 \cdots \\[2mm]
\hat{\alpha}_4 &= \frac{168}{85}\, y^2 - 1.690526\, y^3 + 17.741912\, y^4 \cdots \\[2mm]
\hat{\alpha}_6 &= -0.949967\, y^3 \cdots
\end{aligned}
\tag{8.3}
$$

$$B_{\text{Bar}} = \frac{98}{135}\, y^3 \left(1 - \frac{116}{255}\, y + 2.645997\, y^2 - 0.292781\, y^3 \cdots\right) \tag{8.4}$$

In comparing Eq. (8.2) with Eq. (8.4) one recognizes that only the lowest order coincides due to the fact that quadrupole and hexadecapole deformations are not identical to spheroidal deformations. Eqs.(8.1) and (8.3) are connected by

$$\hat{\alpha}_2 = \frac{1}{3}\hat{e}^2 \left(1 + \frac{20}{21}\hat{e}^2 + \frac{92}{315}\hat{e}^4 \cdots\right)$$

$$\hat{\alpha}_4 = \frac{3}{35}\hat{e}^4 \cdots \tag{8.5}$$

In the natural units of Chap. 1 the other quantities at the $\hat{\alpha}_2, \hat{\alpha}_4$- saddle points are

$$\hat{\varrho}_n = 1 - \frac{7}{6}y\left(1 - \frac{58}{255}y \cdots\right)$$

$$\hat{\zeta}_0 = 1 + \frac{7}{3}y\left(1 - \frac{37}{255}y \cdots\right)$$

$$\hat{Q} = \frac{56}{15}\pi\, y\left(1 + \frac{206}{255}y \cdots\right) \tag{8.6}$$

$$\hat{\mathcal{J}}_{\parallel} = 1 - \frac{7}{3}y\left(1 - \frac{389}{255}y \cdots\right)$$

$$\hat{\mathcal{J}}_{\perp} = 1 + \frac{7}{6}y\left(1 + \frac{1396}{255}y \cdots\right)$$

$$\hat{\mathcal{J}}_{\mathrm{eff}}^{-1} = \frac{7}{2}y\left(1 + \frac{206}{255}y \cdots\right) \tag{8.7}$$

$$\hat{B}_{\mathrm{Coul}} = 1 - \frac{49}{45}y^2\left(1 - \frac{464}{765}y \cdots\right)$$

$$\hat{B}_{\mathrm{surf}} = 1 + \frac{98}{45}y^2\left(1 - \frac{974}{765}y \cdots\right) \tag{8.8}$$

$$\hat{B}_{\mathrm{curv}} = 1 + \frac{98}{45}y^2\left(1 - \frac{124}{765}y \cdots\right)$$

$$\hat{B}_{\mathrm{comp}} = 1 + \frac{196}{45}y^2\left(1 - \frac{974}{765}y \cdots\right) \tag{8.9}$$

$$\hat{B}_{\mathrm{red}} = 1 + \frac{98}{45}y^2\left(1 - \frac{124}{765}y \cdots\right)$$

$$\hat{B}_{\mathrm{sr1}} = 1 - \frac{98}{45}y^2\left(1 - \frac{622}{765}y \cdots\right)$$

$$\hat{B}_{\mathrm{sr2}} = 1 + 0\, y^2 \cdots \tag{8.10}$$

Including the rotational energy, for small values $z \ll 1$ of the rotational parameter, cf. Eq. (1.78), the saddle point deformations become [Bohr 75]

$$\hat{\alpha}_2 = \frac{7}{3}\left(y - \frac{15}{56}\frac{z}{y} \cdots\right)$$

$$\hat{\gamma} = \frac{5\sqrt{3}}{56}\frac{z}{y^2} \cdots \tag{8.11}$$

and the relative barrier energy is lowered to

$$B_{\mathrm{Bar}} = \frac{98}{135}\left(y^3 - \frac{45}{28}yz \cdots\right) \tag{8.12}$$

For larger angular momenta one has

$$\hat{\alpha}_2 = \frac{7}{6} y \left(4 - \frac{15z}{7y^2}\right)^{-1/2}$$

$$\hat{\gamma} = \frac{\pi}{3} - \arccos\left(4 - \frac{15z}{7y^2}\right)^{-1/2} . \tag{8.13}$$

Then the ground state deformations given by Eq. (6.103) with $\alpha_B = \alpha_2$ and the saddle point Eq. (8.11) merge into each other and at the critical value

$$z^{\mathrm{crit}} = \frac{7}{5} y^2 , \tag{8.14}$$

where

$$\alpha_2^{\mathrm{crit}} = \frac{7}{6} y$$

$$\gamma^{\mathrm{crit}} = \frac{\pi}{3}, \tag{8.15}$$

beyond which no stable equilibrium exists. Thorough studies on the rotating liquid drop can be found in [Chandrasekhar 61], [Beringer 61], [Cohen 74], [Pik-Pichak 80], [Brown 80].

Although Eq. (8.4) is based on an expansion in $y \ll 1$ it is accurate within 10% down to $x \approx 0.7$ [Swiatecki 56a]. On the other hand, for $x \ll 1$ one obtains for a configuration of two tangent spheres, cf. Chap. 9,

$$B_{\mathrm{Bar}} = (2^{1/3} - 1) + 2x \left(2^{-2/3} + \frac{5}{24} 2^{1/3} - 1\right)$$

$$= 0.25992 - 0.21511\, x. \tag{8.16}$$

Interpolations for all fissilities based on numerical studies [Swiatecki 56a], [Cohen 63] yield

$$B_{\mathrm{Bar}} \approx \begin{cases} 0.38\left(\dfrac{3}{4} - x\right) & , \ \frac{1}{3} \leq x \leq \frac{2}{3} \\[2mm] 0.83\,(1 - x)^3 & , \ \frac{2}{3} \leq x \leq 1, \end{cases} \tag{8.17}$$

or, better

$$B_{\mathrm{Bar}} \approx \begin{cases} 0.2599 - 0.2151x - 0.1643x^2 - 0.0673x^3 & , \ 0 \leq x \leq 0.6 \\[2mm] 0.7259y^3 - 0.3302y^4 + 0.6387y^5 \\ +7.8727y^6 - 12.0061y^7 & , \ 0.6 \leq x \leq 1. \end{cases} \tag{8.18}$$

Fig. 8.1 shows the relative barrier energy obtained from an accurate numerical calculation based on the matched quadratic surfaces parameterization of Sect. 7.4.3 and Fig. 8.2 displays the corresponding shapes.

8.2 Businaro-Gallone Point

The stability of the saddle point with respect to asymmetric deformations has been studied by [Businaro 55a], [Businaro 55b], [Businaro 57]. Suppose a nearly spherical shape with fissility $x \approx 1$, i.e. $y \ll 1$ and using the α_n-

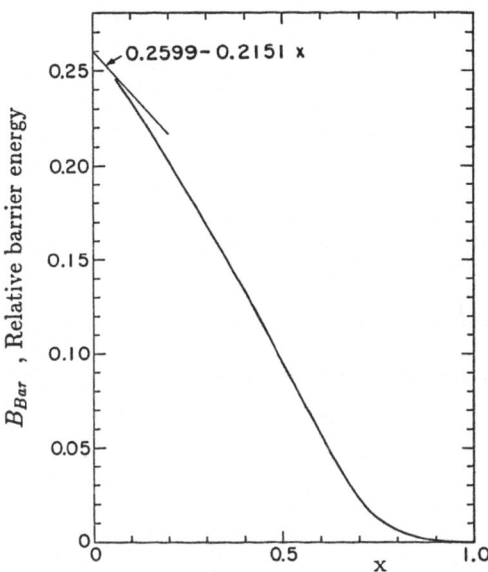

Figure 8.1: The relative barrier energy in the liquid drop model (from [Nix 69]).

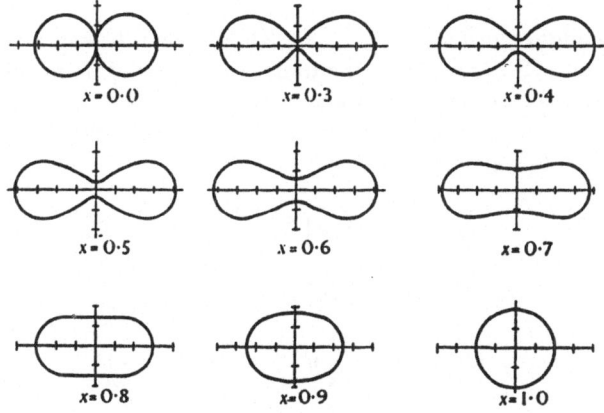

Figure 8.2: Various liquid drop saddle point shapes (from [Cohen 63]).

parameterization. Then the deformation energy is minimized with respect to α_5 to yield

$$\hat{\alpha}_5 = \frac{1250}{2793}\,\hat{\alpha}_2\hat{\alpha}_3\cdots \qquad (8.19)$$

The term in the relative deformation energy proportional to $\hat{\alpha}_3$ becomes negative at

$$x_{\mathrm{BG}}^{\mathrm{sphere}} = 0.7540\ , \qquad (8.20)$$

below which the saddle point becomes instable with respect to asymmetric distortions. This value, however, is not close to unity. Using slightly asymmetric tangent spheres, on the other hand, yields

$$x_{\text{BG}}^{\text{two spheres}} = 0.6 \qquad (8.21)$$

exactly, which is not close to one. Numerical studies [Cohen 63], [Nix 69], [Hasse 71] with different shape parameterizations give

$$x_{\text{BG}} \approx \begin{cases} 0.394 & , \text{Cohen} \\ 0.396 & , \text{Nix} \\ 0.422 & , \text{Hasse} . \end{cases} \qquad (8.22)$$

8.3 Normal Modes

Using the spheroidal eccentricity e as parameter the stiffness, hydrodynamic mass parameter (inertia) and squared normal frequency of Sect. 1.8 at the saddle point for $y \ll 1$ become

$$\begin{aligned}
\hat{C}_e &= -\frac{112}{45} y^2 \left(1 + \frac{32}{3} y \cdots\right) \\
\hat{B}_e &= \frac{14}{15} y \left(1 + \frac{46}{3} y \cdots\right) \\
\hat{\omega}_e^2 &= -\frac{8}{3} y \left(1 - \frac{14}{3} y \cdots\right)
\end{aligned} \qquad (8.23)$$

Note that the higher order terms in Eq. (8.23) have the same poor quality as those in (8.1) and (8.2). Since in one dimension the saddle point is a maximum the stiffness and the squared frequency are negative.

In the α_n- parameterization with $n = 2, 3, 4$, on the other hand, stiffness and mass parameters are given by

$$\begin{aligned}
\hat{C}_{22} &= -\frac{4}{5} y \left(1 - \frac{848}{255} y \cdots\right) \\
\hat{C}_{44} &= \frac{34}{27} \left(1 + \frac{10}{7} y \cdots\right) \\
\hat{C}_{24} &= -\frac{32}{15} y \left(1 - \frac{1301}{1020} y \cdots\right) \\
\hat{C}_{33} &= \frac{30}{49} \left(1 - \frac{52}{45} y \cdots\right)
\end{aligned} \qquad (8.24)$$

$$\begin{aligned}
\hat{B}_{22} &= \frac{3}{10} (1 + 3y \cdots) \\
\hat{B}_{44} &= \frac{1}{12} \left(1 + \frac{25}{11} y \cdots\right) \\
\hat{B}_{24} &= \frac{1}{2} y \cdots \\
\hat{B}_{33} &= \frac{1}{7} \left(1 + \frac{112}{45} y \cdots\right)
\end{aligned} \qquad (8.25)$$

Note that the coupling terms between even and odd parameters vanish. The normal modes at the saddle point consist of an imaginary fission mode ω_F and a vibrational mode ω_V which are coupled to $\hat{\alpha}_2$ and $\hat{\alpha}_4$ and a decoupled asymmetry-vibration mode ω_3,

$$\hat{\omega}_F^2 = -\frac{8}{3} y \left(1 + \frac{304}{255} y \cdots \right)$$

$$\hat{\omega}_V^2 = \frac{136}{9} \left(1 - \frac{515}{187} y \cdots \right)$$

$$\hat{\omega}_3^2 = \frac{30}{7} \left(1 - \frac{164}{45} y \cdots \right). \tag{8.26}$$

Chapter 9

Separated Shapes

9.1 Two Spheres

Qualitative calculations of fission or heavy ion scattering properties often involve a configuration of two tangent spheres. In simplifying the dumbbell-parameterization of Sect. 7.4.2 one arrives at the one-parameter family of shapes

$$\varrho_s^2 = \begin{cases} -z(z + 2r_1) & , \ z \le 0 \\ -z(z - 2r_2) & , \ z \ge 0 \ , \end{cases} \tag{9.1}$$

where r_1, r_2 are the radii of the *left* and *right* spheres. Let $\mu = r_1/r_2$ be the ratio of radii (i.e. μ^3 is the volume ratio) then the expressions

$$\begin{aligned} r_1 &= R_0(1 + \mu^3)^{-1/3} \\ r_2 &= R_0(1 + \mu^{-3})^{-1/3} \end{aligned} \tag{9.2}$$

result from volume conservation of the compound system, $V = 4\pi R_0^3/3$.

In the natural units of Chap. 1 the geometrical quantities read

$$\begin{aligned} \zeta_0 &= \frac{1 + \mu}{(1 + \mu^3)^{1/3}} \\[2mm] Q &= \frac{8}{3}\pi \frac{1 + \mu^5}{(1 + \mu^3)^{5/3}} \\[2mm] r_{ms}^2 &= \frac{8}{5} \frac{1 + \mu^5}{(1 + \mu^3)^{5/3}} \\[2mm] \mathcal{J}_\parallel &= \frac{1 + \mu^5}{(1 + \mu^3)^{5/3}} \\[2mm] \mathcal{J}_\perp &= \frac{7}{2} \frac{1 + \mu^5}{(1 + \mu^3)^{5/3}} \\[2mm] \mathcal{J}_{eff}^{-1} &= \frac{5}{7} \frac{(1 + \mu^3)^{5/3}}{1 + \mu^5} \end{aligned} \tag{9.3}$$
$$\tag{9.4}$$

and the relative Coulomb-self energy and relative Coulomb-interaction energy for tangent spheres become, cf. [Ryce 72], [Ryce 65],

$$B_{Coul,\ self} = \frac{1 + \mu^5}{(1 + \mu^3)^{5/3}} \tag{9.5}$$

$$B_{\text{Coul, int}} = \frac{5}{3} \frac{\mu^3}{(1+\mu)(1+\mu^3)^{5/3}} \tag{9.6}$$

and the other relative energies described in Chap. 1 are

$$B_{\text{surf}} = \frac{1+\mu^2}{(1+\mu^3)^{2/3}}$$

$$B_{\text{curv}} = \frac{1+\mu}{(1+\mu^3)^{1/3}}$$

$$B_{\text{comp}} = \frac{(1+\mu^2)^2}{(1+\mu^3)^{4/3}} \tag{9.7}$$

$$B_{\text{red}} = \frac{175}{2}\left[F(\mu) + F\left(\frac{1}{\mu}\right) - \frac{24}{25}B_{\text{Coul}}^2\right]$$

with $F(\mu) = (1+\mu^3)^{-7/3}$

$$\times \left\{\frac{34}{35} + \mu^6\left[1 - \frac{\mu(\mu+2)}{2(\mu+1)}\log\left(1+\frac{2}{\mu}\right)\right] + \frac{8\mu^3}{5(1+\mu)}\right\} \tag{9.8}$$

$$B_{\text{sr1}} = (5\overline{\Phi}_s)^2 - 60\overline{\Phi}_s B_{\text{Coul}} B_{\text{surf}} + (6B_{\text{Coul}}B_{\text{surf}})^2$$

$$\overline{\Phi}_s = G(\mu) + G\left(\frac{1}{\mu}\right),$$

$$G(\mu) = \frac{1+\mu+\mu^3}{(1+\mu)(1+\mu^3)^{4/3}} \tag{9.9}$$

$$B_{\text{sr2}} = 25\overline{\Phi_s^2} - 60\overline{\Phi}_s B_{\text{Coul}} + 36B_{\text{Coul}}^2 B_{\text{surf}}$$

$$\overline{\Phi_s^2} = H(\mu) + H\left(\frac{1}{\mu}\right),$$

with $H(\mu) = (1+\mu^3)^{-2}$

$$\times \left[1 + \frac{2\mu^3}{1+\mu} + \frac{\mu^6}{2(1+\mu)}\log\left(1+\frac{2}{\mu}\right)\right]. \tag{9.10}$$

The relative deformation energy of the system characterized by the fissility x of the compound nucleus as a function of the ratio of radii is shown in Fig. 9.1. It is used, for instance, to estimate the direction of mass drift of a heavy-ion system at contact toward asymmetry or symmetry, the transition occurring at $x = 0.6$.

For slightly asymmetric systems we can define the small quantity ν, where $\mu = 1 + \nu$. Then the geometrical quantities above become

$$\zeta_0 = 2^{2/3}\left(1 - \frac{1}{4}\nu^2\cdots\right)$$

$$Q = \frac{4}{3}\pi 2^{1/3}\left(1 + \frac{15}{4}\nu^2\cdots\right)$$

$$r_{\text{ms}}^2 = \frac{4}{5}2^{1/3}\left(1 + \frac{15}{4}\nu^2\cdots\right) \tag{9.11}$$

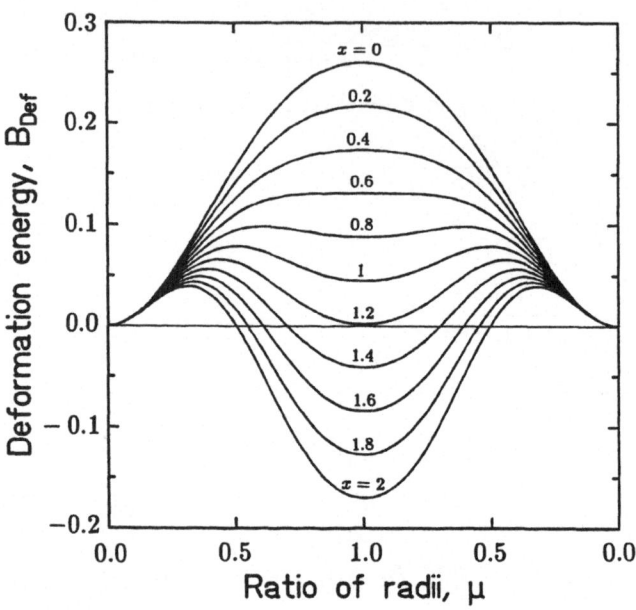

Figure 9.1: The deformation energy of a system of two tangent spheres as a function of the ratio of radii for various fissilities.

$$\mathcal{J}_{\parallel} = \frac{1}{2} 2^{1/3} \left(1 + \frac{15}{4} \nu^2 \cdots \right)$$

$$\mathcal{J}_{\perp} = \frac{7}{4} 2^{1/3} \left(1 + \frac{15}{4} \nu^2 \cdots \right)$$

$$\mathcal{J}_{\text{eff}}^{-1} = \frac{7}{4} 2^{2/3} \left(1 + \frac{15}{4} \nu^2 \cdots \right) \tag{9.12}$$

$$B_{\text{Coul,self}} = \frac{1}{2} 2^{1/3} \left(1 + \frac{5}{4} \nu^2 \cdots \right)$$

$$B_{\text{Coul,int}} = \frac{5}{24} 2^{1/3} \left(1 - 2\nu^2 \cdots \right) \tag{9.13}$$

$$B_{\text{surf}} = 2^{1/3} \left(1 - \frac{1}{4} \nu^2 \cdots \right)$$

$$B_{\text{curv}} = 2^{2/3} \left(1 - \frac{1}{4} \nu^2 \cdots \right)$$

$$B_{\text{comp}} = 2^{2/3} \left(1 - \frac{1}{2} \nu^2 \cdots \right) \tag{9.14}$$

$$
\begin{aligned}
B_{\text{red}} &= \frac{175}{16} 2^{2/3} \left(\frac{887}{525} - \frac{3}{2} \log 3 \right) \left(1 + \frac{21420 - 18375 \log 3}{7096 - 6300 \log 3} \nu^2 \cdots \right) \\
&= 0.72236(1 + 7.05609\, \nu^2 \cdots) \tag{9.15}
\end{aligned}
$$

108

$$B_{\text{sr1}} = \frac{1}{2} 2^{1/3} \left(1 + \frac{13}{4}\nu^2 \cdots\right)$$

$$B_{\text{sr2}} = \frac{1}{8}(25 \log 3 - 21)\left(1 + \frac{675 \log 3 - 221}{900 \log 3 - 756}\nu^2 \cdots\right)$$

$$= 0.80816(1 + 2.23657\nu^2 \cdots) \qquad (9.16)$$

The interaction part of the Krappe-Nix finite range surface energy of Sect. 1.4.2 for two separated or tangent spheres with distance $a\delta$ between their tips reads

$$B_{\text{KN,int}} = -\frac{4}{\sigma^2}(\sigma_1 \cosh \sigma_1 - \sinh \sigma_1)(\sigma_2 \cosh \sigma_2 - \sinh \sigma_2)\frac{e^{-\lambda}}{\lambda}, \qquad (9.17)$$

where $\sigma = R_0/a$, $\sigma_1 = R_1/a$, $\sigma_2 = R_2/a$ and

$$\lambda = \delta + \sigma_1 + \sigma_2 \qquad (9.18)$$

is the distance between the centers of mass in units of a. By volume conservation, $\sigma^3 = \sigma_1^3 + \sigma_2^3$. For $\sigma_1, \sigma_2 \gg 1$ this reduces to

$$B_{\text{KN,int}} = -\frac{\sigma_1 \sigma_2}{\sigma^2(\sigma_1 + \sigma_2)}e^{-\delta}. \qquad (9.19)$$

Similarly,

$$B_{\text{YE,int}} = -\frac{4}{\sigma^2 \lambda}g(\sigma_1)g(\sigma_2)\left(4 + \lambda - \frac{f(\sigma_1)}{g(\sigma_1)} - \frac{f(\sigma_2)}{g(\sigma_2)}\right)e^{-\lambda}$$

$$f(x) = x^2 \sinh x$$

$$g(x) = x \cosh x - \sinh x . \qquad (9.20)$$

Neglecting terms of the order of $e^{-\sigma_1}$, $e^{-\sigma_2}$ this simplifies to

$$B_{\text{YE,int}} = -\frac{(\sigma_1 - 1)(\sigma_2 - 1)}{\sigma^2 \lambda}\left(4 + \lambda - \frac{\sigma_1}{\sigma_1 - 1} - \frac{\sigma_2}{\sigma_2 - 1}\right)e^{-\delta}. \qquad (9.21)$$

For two cylindrically symmetric objects with parabolic crevices of radii R_1, R_2 the proximity energy simplifies to

$$E_{\text{prox}} = E_s^0 \frac{\overline{R}a}{R_0^2} \Phi(\delta), \qquad (9.22)$$

where

$$\overline{R}^{-1} = R_1^{-1} + R_2^{-1} \qquad (9.23)$$

is the reduced radius and

$$\Phi(x) = \int_x^\infty dx \, \varphi(x) \qquad (9.24)$$

is the integral of the universal function φ; R_0 is the radius of the compound nucleus and $d = a\delta$ is the distance between the half density radii which might be negative for overlapping configurations.

Under the assumption that the curvature radii of the crevices coincide with the nuclear radii the relative proximity energy in the same notation as above

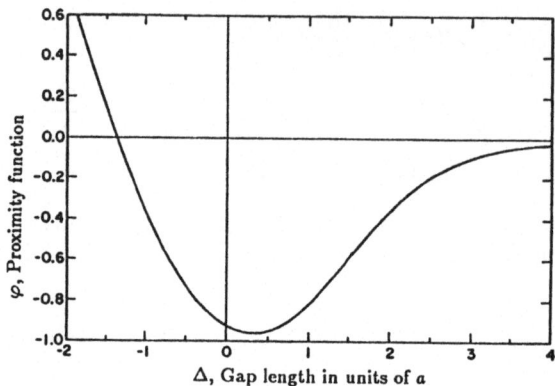

Figure 9.2: The universal proximity function as a function of the gap length in units of a (from [Błocki 77]).

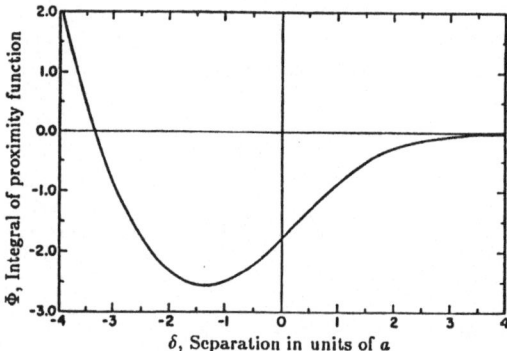

Figure 9.3: The first integral of the universal proximity function as a function of the distance between the surfaces in units of a. (from [Błocki 77]).

becomes

$$B_{\text{prox}} = \frac{\sigma_1 \sigma_2}{\sigma^2(\sigma_1 + \sigma_2)} \, \Phi(\delta). \tag{9.25}$$

The universal proximity function and its first integral are shown in Figs. 9.2 and 9.3. The latter can be approximated by

$$\Phi(\delta) \approx \begin{cases} -0.5(\delta - 2.54)^2 - 0.0852(\delta - 2.54)^3 & , \ 0 < \delta < 1.2511 \\ -3.437 e^{-\delta/0.75} & , \ \delta > 1.2511 \end{cases} \tag{9.26}$$

or, better [Błocki 81],

$$\Phi(\delta) \approx \begin{cases} -1.7817 + \delta & , \ \delta < 0 \ (d) \\ -1.7817 + 0.9270\delta + 0.143\delta^2 - 0.09\delta^3 & , \ \delta < 0 \ (s) \\ -1.7817 + 0.9270\delta + 0.1696\delta^2 - 0.05148\delta^3 & , \ 0 < \delta < 1.9475 \\ -4.41 e^{-\delta/0.7176} \ (\text{exact for } \delta > 2.74) & , \ \delta > 1.9475 \ , \end{cases} \tag{9.27}$$

where (d) corresponds to overlapping systems with double density and without volume conservation and (s) with single density with volume conservation. The universal function itself can be obtained by differentiation, $\varphi = -\Phi'$. If the approximation of parabolic crevices is not sufficient, B_{prox} is to be calculated numerically with the help of Eq. (1.54).

In order to cancel the surface energy of two adjacent flat surfaces exactly, i.e. $\varphi(\Delta = 0) = -1$ and $\varphi'(\Delta = 0) = 0$, Feldmeier [Feldmeier 80] modified the universal function itself slightly according to

$$\varphi(\Delta) \approx \begin{cases} -1 + 0.1899\Delta^2 & , \ 0 < \Delta < 1.2311 \\[2mm] \begin{aligned} -0.135 - 0.1881\eta - 0.11581\eta^2 \\ -0.01202\eta^3 + 0.02055\eta^4 \end{aligned} & , \ 1.2311 < \Delta < 2.74 \\[2mm] -6.145e^{-\Delta/0.7176} & , \ \Delta > 2.74 \end{cases} \qquad (9.28)$$

with $\eta = 2.74 - \Delta$ but used the full expression Eq. (1.54). Also, the definition of separation distance has been modified according to [Błocki 81]. Rather than using $d = a\delta$,

$$\delta = \lambda - \sigma_1 - \sigma_2 + \frac{1}{2\sigma_1} + \frac{1}{2\sigma_2} \qquad (9.29)$$

is employed where $\ell = a\lambda$ is the better defined distance between the centers of mass of the fragments. An extension of the proximity model to deformed nuclei can be found in [Baltz 82].

9.2 Two Spheroids

Here and in the following chapter formulas are given for the Coulomb interaction energies of two separated or tangent fragments. Surface and Coulomb-self energies are not considered because they are easily derived from the contents of the previous chapters.

The most general but rarely used cases of two arbitrarily oriented spheroids can be found in [Hirschfelder 54] and [Nix 64]. If we restrict ourselves to coplanar spheroids oriented by the same angle θ with respect to the axis joining their centers according to Fig. 9.4, the system is specified by the semiaxes $c_{(1)}$, $a_{(1)}$, $c_{(2)}$, $a_{(2)}$, the angle θ and the distance l between the centers*.

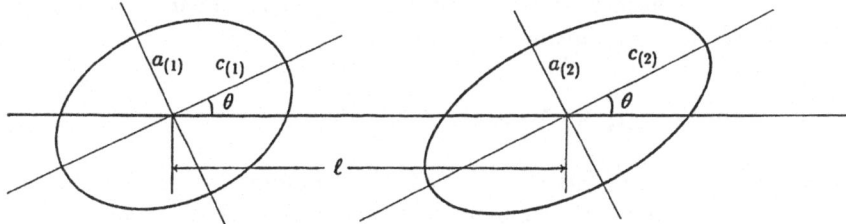

Figure 9.4: The parameters employed for describing two coplanar spheroids.

---

*Here and in the following indices in parentheses refer to fragment numbers

With the abbreviations $(i = 1, 2)$,

$$e_{(i)}^2 = \begin{cases} \dfrac{c_{(i)}^2 - a_{(i)}^2}{l^2} & \text{, prolate} \\[2ex] \dfrac{a_{(i)}^2 - c_{(i)}^2}{l^2} & \text{, oblate} \end{cases} \tag{9.30}$$

the Coulomb interaction energy of two homogeneously charged spheroids reads

$$E_{\text{Coul,int}} = \frac{q_{(1)}q_{(2)}}{l} \left[s(e_{(1)}, \theta) + s(e_{(2)}, \theta) - 1 + S(e_{(1)}, e_{(2)}, \theta) \right], \tag{9.31}$$

where $q_{(i)}$ are the charges of the respective spheroids and

$$s(e_{(i)}, \theta) = 3 \sum_{n=0}^{\infty} \frac{P_{2n}(\cos\theta)}{(2n+1)(2n+3)} e_{(i)}^{2n}$$

$$S(e_{(1)}, e_{(2)}, \theta) = 9 \sum_{j,k=1}^{\infty} \frac{(2j+2k)! \, P_{2j+2k}(\cos\theta)}{(2j+1)(2j+3)(2k+1)(2k+3)(2j)! \, (2k)!} e_{(1)}^{2j} e_{(2)}^{2k}. \tag{9.32}$$

For collinear spheroids $(\theta = 0)$ one has $P_l(\cos\theta) = 1$ and

$$s(e, \theta) = \begin{cases} \dfrac{3}{4} \left(\dfrac{1}{e} - \dfrac{1}{e^3} \right) \log \dfrac{1+e}{1-e} + \dfrac{3}{2e^2} & \text{, prolate} \\[2ex] \dfrac{3}{2} \left(\dfrac{1}{e} + \dfrac{1}{e^3} \right) \arctan e - \dfrac{3}{2e^2} & \text{, oblate.} \end{cases} \tag{9.33}$$

In this case the whole series can be summed up to give [Quentin 69], cf. also [Schultheis 75],

$$E_{\text{Coul,int}} = \frac{q_{(1)}q_{(2)}}{l} \frac{3}{40} \left(\frac{1 + 11e_{(1)}^2 + 11e_{(2)}^2}{e_{(1)}^2 \, e_{(2)}^2} + T(e_{(1)}, e_{(2)}) \right)$$

$$T(e_{(1)}, e_{(2)}) = \mathcal{P} \Big\{ \frac{(1 + e_{(1)} + e_{(2)})^3}{e_{(1)}^3 \, e_{(2)}^3} [\log(1 + e_{(1)} + e_{(2)})]$$

$$\times (1 - 3e_{(1)} - 3e_{(2)} + 12e_{(1)}e_{(2)} - 4e_{(1)}^2 - 4e_{(2)}^2) \Big\}. \tag{9.34}$$

Here $\mathcal{P}$ denotes the even part in $e_{(1)}$ and $e_{(2)}$, i.e. (9.34) must be expanded in a series and only the even terms retained.

9.3 Two Distorted Spheres

By using the methods of [Hirschfelder 54] the Coulomb-interaction energy of two homogeneously charged fragments in the α_n- parameterization reads

$$E_{\text{Coul,int}}^{\text{hom}} = \frac{q_{(1)} \, q_{(2)}}{\lambda_{(1)}^3 \, \lambda_{(2)}^3} \sum_{m,n=0}^{\infty} \frac{(m+n)!}{m! \, n!} \frac{q_{(1)m} \, q_{(2)n}}{\tilde{l}_{(1)}^m \, \tilde{l}_{(2)}^n}, \tag{9.35}$$

where $\tilde{l}_{(i)} = \lambda_{(i)} l/R_0$ and l is the distance between the two centers. The quantities $q_{(i)k}$ are the reduced multipoles of Sect. 6.2.1 for the respective fragment.

Up to quadratic order in $\alpha_{(i)n}$ and sixth order in l^{-1} this becomes

$$
\begin{aligned}
E_{\text{Coul,int}}^{\text{hom}} = \frac{q_{(1)}q_{(2)}}{l} \Bigg\{ &1 + l^{-2} \left[R_{(1)0}^2 \left(\frac{3}{5}\alpha_{(1)2} + \frac{12}{35}\alpha_{(1)2}^2 + \frac{8}{35}\alpha_{(1)3}^2 \right) \right.\\
&+ R_{(2)0}^2 \left(\frac{3}{5}\alpha_{(2)2} + \frac{12}{35}\alpha_{(2)2}^2 + \frac{8}{35}\alpha_{(2)3}^2 \right) \Bigg] + l^{-3} \left[R_{(1)0}^3 \left(\frac{3}{7}\alpha_{(1)3} \right.\right.\\
&+ \frac{4}{7}\alpha_{(1)2}\alpha_{(2)3} \right) + R_{(2)0}^3 \left(\frac{3}{7}\alpha_{(2)3} + \frac{4}{7}\alpha_{(2)2}\alpha_{(2)3} \right) \Bigg]\\
&+ l^{-4} \left[R_{(1)0}^4 \left(\frac{18}{35}\alpha_{(1)2}^2 + \frac{18}{77}\alpha_{(1)3}^2 \right) + R_{(2)0}^4 \left(\frac{18}{35}\alpha_{(2)2}^2 + \frac{18}{77}\alpha_{(2)3}^2 \right) \right.\\
&+ R_{(1)0}^2 R_{(2)0}^2 \frac{54}{25}\alpha_{(1)2}\alpha_{(2)2} \Bigg] + l^{-5} \left[R_{(1)0}^5 \frac{10}{11}\alpha_{(1)2}\alpha_{(1)3} \right.\\
&+ R_{(2)0}^5 \frac{10}{11}\alpha_{(2)2}\alpha_{(2)3} + R_{(1)0}^3 R_{(2)0}^2 \frac{18}{7}\alpha_{(1)3}\alpha_{(2)2}\\
&+ R_{(1)0}^2 R_{(2)0}^3 \frac{18}{7}\alpha_{(1)2}\alpha_{(2)3} \Bigg] + l^{-6} \left[R_{(1)0}^6 \frac{400}{1001}\alpha_{(1)3}^2 \right.\\
&+ R_{(2)0}^6 \frac{400}{1001}\alpha_{(2)3}^2 + R_{(1)0}^3 R_{(2)0}^3 \frac{180}{49}\alpha_{(1)3}\alpha_{(2)3} \Bigg] \Bigg\} \cdots
\end{aligned}
\tag{9.36}
$$

For two equal fragments and $\gamma = l/R_0$, it simplifies to

$$
\begin{aligned}
B_{\text{Coul,int}}^{\text{hom}} = 2 \Bigg[&\frac{5}{6}\gamma^{-1} + \gamma^{-3}\left(\alpha_2 + \frac{4}{7}\alpha_2^2 + \frac{8}{21}\alpha_3^2 \right)\\
&+ \gamma^{-4}\alpha_3\left(\frac{5}{7} + \frac{20}{21}\alpha_2 \right) + \gamma^{-5}\left(\frac{93}{35}\alpha_2^2 + \frac{30}{77}\alpha_3^2 \right)\\
&+ \gamma^{-6}\frac{1340}{231}\alpha_2\alpha_3 + \gamma^{-7}\frac{78350}{21021}\alpha_3^2 \Bigg].
\end{aligned}
\tag{9.37}
$$

If the nuclei are inhomogeneously charged according to the convention of Fig. 9.5 and with the charge density of Sect. 6.2.1 one merely has to make the following replacement

$$
q_{(i)m} \to q_{(i)m} - \varepsilon\, t_{(i)m}/\lambda_{(i)}.
\tag{9.38}
$$

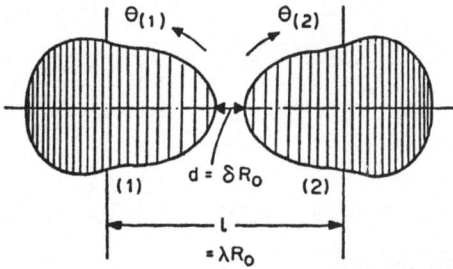

Figure 9.5: Convention of measuring the polar angles θ in order to achieve positive odd shape parameters α_3 and dipole moment ε. The charge densities are indicated by hatching (from [Hasse 78]).

For two equal fragments the inhomogeneous part of the Coulomb interaction energy becomes

$$B_{\text{Coul,int}}^{\text{inhom}} = 2\left\{-\varepsilon\left[\gamma^{-2}\left(\frac{1}{3} + \frac{2}{3}\alpha_2 + \frac{5}{7}\alpha_2^2 + \frac{31}{63}\alpha_2^2\right)\right.\right.$$

$$+\gamma^{-3}\alpha_3\left(\frac{3}{7} + \frac{32}{35}\alpha_2\right) + \gamma^{-4}\left(\frac{36}{35}\alpha_2 + \frac{84}{35}\alpha_2^2 + \frac{268}{385}\alpha_3^2\right)$$

$$+\gamma^{-5}\alpha_3\left(\frac{56}{63} + \frac{16916}{3465}\alpha_2\right) + \gamma^{-6}\left(\frac{304}{77}\alpha_2^2 + \frac{19800}{7007}\alpha_3^2\right)$$

$$+\gamma^{-7}\frac{64800}{7007}\alpha_2\alpha_3 + \gamma^{-8}\frac{54800}{9009}\alpha_3^2\bigg]$$

$$+\varepsilon^2\left[\gamma^{-3}\left(\frac{1}{15} + \frac{4}{15}\alpha_2 + \frac{58}{105}\alpha_2^2 + \frac{62}{315}\alpha_3^2\right) + \gamma^{-4}\alpha_3\left(\frac{9}{35} + \frac{186}{175}\alpha_2\right)\right.$$

$$+\gamma^{-5}\left(\frac{12}{35}\alpha_2 + \frac{48}{35}\alpha_2^2 + \frac{1899}{2695}\alpha_3^2\right) + \gamma^{-6}\alpha_3\left(\frac{20}{63} + \frac{5128}{1617}\alpha_2\right)$$

$$+\gamma^{-7}\left(\frac{930}{539}\alpha_2^2 + \frac{13620}{7007}\alpha_3^2\right) + \gamma^{-8}\frac{4120}{1001}\alpha_2\alpha_3 + \gamma^{-9}\frac{74000}{27027}\alpha_3^2\bigg]\bigg\}$$

$$\cdots$$

(9.39)

Some of the above results for the Coulomb interaction energy are also given in [Geilikman 55], [Geilikman 58], [Geilikman 58a], however, with discrepancies in some terms.

9.4 Three Fragments

Diehl and Greiner [Diehl 74] studied fission of a liquid drop into three fragments in a restricted family of three collinearly aligned or triangle-like configuration, see Fig. 9.6. No simple formulas, however, are available for the parameterization or for the liquid drop energies.

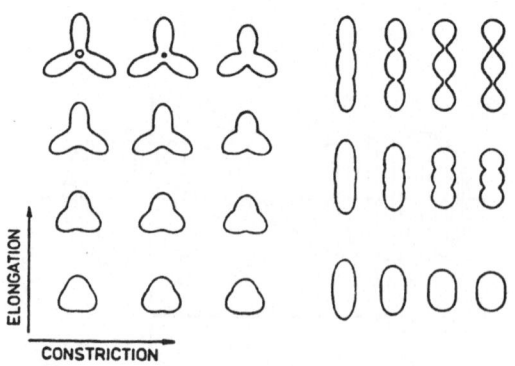

Figure 9.6: Ternary fission shapes (from [Diehl 74]).

9.5 n-Spheres

The energies of a configuration of n infinitely separated equal spheres are needed, for instance, for the question whether spallation into many fragments is exothermal or endothermal. From the shape dependences

$$
\begin{aligned}
B_{\text{surf}} &= n^{1/3} \\
B_{\text{Coul}} &= n^{-2/3} \\
B_{\text{curv}} &= n^{2/3} \\
B_{\text{comp}} &= n^{2/3} \\
B_{\text{red}} &= n^{-4/3} \\
B_{\text{sr1}} &= n^{-2/3} \\
B_{\text{sr2}} &= n^{-1}
\end{aligned}
\tag{9.40}
$$

$$
\tag{9.41}
$$

one finds that for given multiplicity n the critical fissility is

$$
x > \frac{n^{1/3} - 1}{2(1 - n^{-1/3})}
\tag{9.42}
$$

for which spallation is exothermal. From Eq. (9.42) one finds that ordinary binary fission is exothermal for $x > 0.351$ and that for $x = 1$ the largest number of fragments that result in an energy release is 20, see Fig. 9.7.

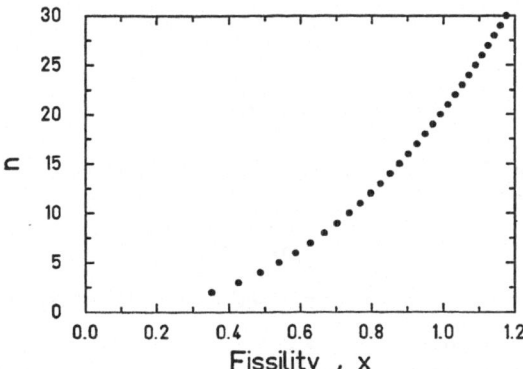

Figure 9.7: The maximum number of fragments n for which fission of a nucleus, characterized by the fissility x, is exothermal.

Chapter 10

Exotic Shapes

Toroidal and bubble nuclei were investigated in the context of the existence of superheavy nuclei [Wong 73], [Wong 78].

10.1 Toroids

In plane toroidal coordinates (7.86) the shape is defined as the locus of constant η, say η_0. Then the radius w of the orifice is expressed as

$$w = \sqrt{r_t^2 - r_s^2} \tag{10.1}$$

in terms of the toroid (major) radius r_t and the sausage (minor) radius r_s. The latter are eliminated by volume conservation

$$
\begin{aligned}
V &= \frac{4}{3}\pi R_0^3 = 2\pi^2 r_t r_s^2 \\
r_t &= R_0 \left(\frac{2}{3\pi} \cosh^2 \eta_0 \right)^{1/3} \\
r_s &= R_0 \left(\frac{2}{3\pi} \cosh \eta_0 \right)^{1/3}
\end{aligned}
\tag{10.2}
$$

and the relative surface energy follows from the surface area $4\pi^2 r_t r_s$,

$$B_{\text{surf}} = \left(\frac{4\pi}{9} \cosh \eta_0 \right)^{1/3}. \tag{10.3}$$

The relative Coulomb energy is expressed as an infinite sum,

$$
\begin{aligned}
B_{\text{Coul}} = \frac{5}{6} \left(\frac{3\pi}{2} \right)^{1/3} \frac{\sinh^5 \eta_0}{\cosh^{5/3} \eta_0} \\
\times \left[\frac{8}{9\pi^3} \sum_{n=0}^{\infty} (2 - \delta_{n0}) A_n(\cosh \eta_0) - \frac{3 + 4\cosh^2 \eta_0}{8\pi} \frac{\cosh \eta_0}{\sinh^5 \eta_0} \right],
\end{aligned}
\tag{10.4}
$$

where the coefficients ($x = \cosh \eta_0$) read

$$
\begin{aligned}
A_n(x) = \left[\left(n + \frac{1}{2} \right) P_{n+1/2}(x) Q_{n-1/2}^2(x) - \left(n - \frac{3}{2} \right) P_{n-1/2}(x) Q_{n+1/2}^2(x) \right] \\
\times \left[\left(n + \frac{1}{2} \right) Q_{n+1/2}(x) Q_{n-1/2}^2(x) - \left(n - \frac{3}{2} \right) Q_{n-1/2}(x) Q_{n+1/2}^2(x) \right]
\end{aligned}
\tag{10.5}
$$

and P_ν, Q_ν are Legendre functions of first and second kind, respectively. The

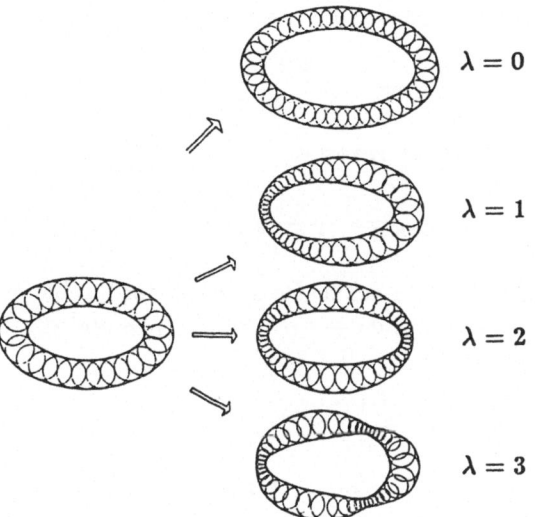

Figure 10.1: Toroids distorted by sausage deformation of order $\lambda =0,1,2,3$. The order corresponds to the maxima or minima in the cross section of the sausage. (After [Wong 73]).

quantity η_0 corresponds to a breathing-type deformation because the toroid remains a toroid. Other types of deformation, cf. Fig. 10.1, are defined by

$$\eta_s = \overline{\eta}_0(1 + \sigma_\lambda \cos \lambda\phi) \quad \lambda = 1, 2, 3, \cdots \tag{10.6}$$

which now varies the cross section depending on the azimuthal angle. $\overline{\eta}_0$ is to be eliminated by volume conservation.

$$\overline{\eta}_0 = \eta_0 + \sigma_\lambda^2 \, \eta_0^2 \, \coth\eta_0 \, \frac{3\coth^2\eta_0 - 2}{3\coth^2\eta_0 - 1}. \tag{10.7}$$

The appropriate formulas for the surface and Coulomb energies can also be found in [Wong 73].

10.2 Bubbles

Bubbles with outer radius r_2 and inner radius r_1 are expressed in terms of the single variable [Wong 73]

$$p = \frac{r_1}{r_2} \leq 1. \tag{10.8}$$

The radii are eliminated by volume conservation

$$\begin{aligned}
V &= \frac{4}{3}\pi R_0^3 = \frac{4}{3}\pi(r_2^3 - r_1^3) \\
r_2 &= R_0(1 - p^3)^{-1/3} \\
r_1 &= R_0\, p(1 - p^3)^{-1/3}.
\end{aligned} \tag{10.9}$$

The relative surface energy follows from the total surface area $4\pi(r_1^2 + r_2^2)$,

$$B_{\text{surf}} = \frac{1 + p^2}{(1 - p^3)^{2/3}} \tag{10.10}$$

and the relative Coulomb energy reads

$$B_{\text{Coul}} = \frac{1 - \frac{5}{2}p^3 + \frac{3}{2}p^5}{(1 - p^3)^{5/3}}. \tag{10.11}$$

The Krappe-Nix energy of Sect. 1.4.2 is also available for bubble nuclei [Krappe 73]. With $\sigma = R_0/a$, $\sigma_1 = r_1/a$, $\sigma_2 = r_2/a$, $p = \sigma_1/\sigma_2$ and $\sigma^3 = \sigma_2^3 - \sigma_1^3$ for volume conservation, it reads

$$\begin{aligned}
B_{\text{KN}} = {}& \frac{2}{3}\sigma - \frac{1}{\sigma^2} - \left(1 + \frac{1}{\sigma}\right)^2 e^{-2\sigma} \\
&+ \frac{1}{\sigma^2}\left[\sigma_1^2 + \sigma_2^2 + \frac{2}{3}\sigma_1^3 - \frac{2}{3}\sigma_2^3 + (1 + \sigma_1)^2 e^{-2\sigma_1} + (1 + \sigma_2)^2 e^{-2\sigma_2}\right. \\
&\left. + 2(1 - \sigma_1)(1 + \sigma_2)e^{-\sigma_2+\sigma_1} - 2(1 + \sigma_1)(1 + \sigma_2)e^{-\sigma_2-\sigma_1}\right]. \tag{10.12}
\end{aligned}$$

Bubble shapes can also be distorted harmonically [Wong 73],

$$R_i(\theta, \phi) = \overline{R}_i\left[1 + \frac{\beta_{i0}}{2\sqrt{\pi}} + \sum_{\lambda>0,\mu} \beta_{i\lambda\mu}\left(Y_{\lambda\mu}(\theta, \phi) + Y_{\lambda\mu}^*(\theta, \phi)\right)\right] \quad i = 1, 2 \tag{10.13}$$

and volume conservation demands

$$\overline{R}_i = r_i\left[1 - \frac{\beta_{i0}^2}{4\pi} - \frac{1}{2\pi}\sum_{\lambda>0,\mu} \beta_{i\lambda\mu}^2\right] \quad i = 1, 2. \tag{10.14}$$

Appropriate formulas for B_{surf}, B_{Coul} can also be found in [Wong 73].

Chapter 11

Medium- and High-Energy Nuclear Collisions

11.1 Factorization

When the energy of a nuclear collision is sufficiently high some aspects of the process such as the energy and angular distributions of the light fragments produced can be treated by assuming straight line trajectories for the first stage of the process (see, for example, work on the fireball model [Gosset 77], the firestreak model [Myers 78], [Gosset 78], [Cecil 80], the rows-on-rows model [Hüfner 77], [Hatch 79], [Knoll 79a], etc.). When such a description is assumed, the asymptotic density distribution for particles of type j in momentum space $F_j(\boldsymbol{p})$ (which is the measured quantity) can be written as

$$F_j(\boldsymbol{p}) = \int \mathrm{d}\boldsymbol{s} \int \mathrm{d}\sigma\, w_b(x,y)\, J_{\boldsymbol{p}'\to\boldsymbol{p}}(\beta_\mathrm{f})\, f_j(\boldsymbol{p}'; \beta_\mathrm{i}, \eta)\,, \qquad (11.1)$$

where

$$\mathrm{d}\boldsymbol{s} = 2\pi b\,\mathrm{d}b \qquad (11.2)$$

is the differential element for summing over the values of the impact parameter b, and

$$\mathrm{d}\sigma = \mathrm{d}x\,\mathrm{d}y \qquad (11.3)$$

is the differential element for summing over the projections of the target and projectile density distributions on the x-y plane normal to the beam direction. The quantity $f_j(\boldsymbol{p}')$ is the final center of mass momentum space density for particles of type j and it depends on the relative velocity of the projectile β_i and the relative numbers of target and projectile particles that are involved. The quantity $J_{\boldsymbol{p}'\to\boldsymbol{p}}$ is the Jakobian transformation from the c.m. frame (the primed frame) to the laboratory frame. It depends only on the laboratory velocity $\beta_\mathrm{f} = v_\mathrm{f}/c$ of the composite system consisting of the subset of target and projectile particles that are being considered. The quantity $w_b(x,y)$ is obtained (for each value of the impact parameter b) by projecting the target and projectile densities on the x-y plane.

11.2 Different Approaches

11.2.1 Fireball Model

The Fireball Model, as it is employed in nuclear physics, is based on the assumption of straight line trajectories for the projectile particles as they cut a

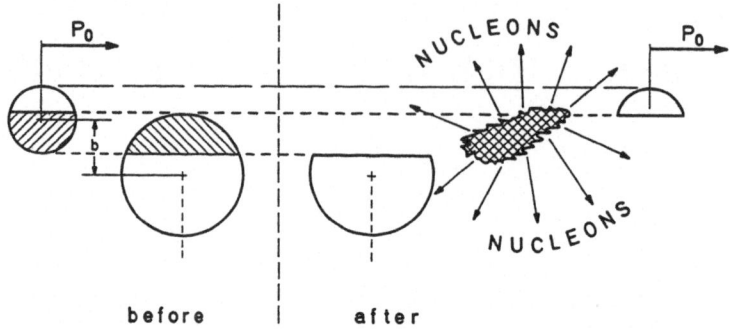

Figure 11.1: An illustration of the straight-line geometry in collisions between nuclei at high energy. An incoming projectile with momentum p_0 per nucleon ploughs through the target. The nucleons from the overlap zone (shaded area) are called *participants*. Only they collide and contribute to the one-nucleon inclusive cross section. The other nucleons (unhatched pieces) remain *spectators*.

cylindrical section out of the target. Similarly, the target is assumed to cut a cylindrical section out of the projectile. (Of course, depending on the relative sizes of the target and projectile and the value of the impact parameter, the entire target or projectile may be swept away). As can be seen in Fig. 11.1 from [Hüfner 77], the swept out portions of the target and projectile are assumed to combine inelastically forming a *participant* fireball. The damaged pieces of the target and projectile (if any) that remain behind are referred to as spectators.

The solution to the problem of calculating the number of participating nucleons of the target and projectile nuclei as a function of impact parameter involves the calculation of the volume of intersection of a sphere and a cylinder. The exact solution to this problem appears to require numerical integrations. An approximate analytical method was developed by Swiatecki and later published in [Gosset 77]. In this approximation the fraction of the nucleus A_1 with radius R_1 that is swept out by a projectile A_2 with radius R_2 and an impact parameter b is given by the functions F below in terms of the quantities,

$$
\begin{aligned}
\nu &= R_1/(R_1 + R_2)\,, \\
\beta &= b/(R_1 + R_2)\,, \\
\mu &= (1 - \nu)/\nu = R_2/R_1\,.
\end{aligned}
\tag{11.4}
$$

If the collision corresponds to a cylindrical hole being cut out of nucleus A_1, with radius R_1

$$
F_I \approx [1 - (1 - \mu^2)^{3/2}]\sqrt{1 - (\beta/\nu)^2}\,.
\tag{11.5}
$$

If a cylindrical channel is cut out of A_1 and $R_1 \geq R_2$ then

$$
F_{II} \approx \frac{3}{4}\sqrt{1 - \nu}\left(\frac{1 - \beta}{\nu}\right)^2
$$
$$
-\frac{1}{8}\left(\frac{3(1 - \nu)^{1/2}}{\mu} - \frac{[1 - (1 - \mu^2)^{3/2}][1 - (1 - \mu)^2]^{1/2}}{\mu^3}\right)\left(\frac{1 - \beta}{\nu}\right)^3.
\tag{11.6}
$$

If a cylindrical channel is cut out of A_1 and $R_1 \leq R_2$ then

$$F_{III} \approx \frac{3}{4}(1-\nu)^{1/2} \left(\frac{1-\beta}{\nu}\right)^2 - \frac{1}{8}\left[3(1-\nu)^{1/2} - 1\right]\left(\frac{1-\beta}{\nu}\right)^3. \tag{11.7}$$

And, in the trivial case where $R_2 \geq R_1$, and A_1 is completely swept away

$$F_{IV} = 1. \tag{11.8}$$

These approximate expressions were compared with numerical calculations in [Gosset 77] and found to be accurate to within a few percent. The worst inaccuracies were for intermediate values of β, the largest being 6% for $\beta = 0.4$ and $\nu = 0.4$

Even though there is no analytic expression for the total number of fireball model participants for a given value of the impact parameter b, the total cross section for producing participant particles can be calculated by re-writing (11.1) as

$$\sigma_{\text{total}} = \int d\boldsymbol{s} \int d\sigma\, q_s(\sigma), \tag{11.9}$$

where q_s here is analogous to $q(\eta)$ below in (11.19). It is a function that determines how many of the particles from the target and projectile that are lined up with each other on straight line trajectories will appear in the final state of interest. Eq. (11.9) can be re-written as

$$\sigma = \int 2\pi s_1\, ds_1 \int 2\pi s_2\, ds_2\, q(s_1, s_2) \tag{11.10}$$

where s_1 and s_2 are one-dimensional radial variables measured from the centers of the projections of the target and projectile density distributions onto a plane. Furthermore, if the nuclei are idealized as spheres of uniform density ϱ and sharp surfaces located at the radii R_1 and R_2, then

$$\sigma = \int\limits_0^{2R_1} \frac{\pi}{2}\alpha\, d\alpha \int\limits_0^{2R_2} \frac{\pi}{2}\beta\, d\beta\, q(\alpha, \beta), \tag{11.11}$$

where α and β are length variables, proportional to the number of particles per unit area when the nuclear densities are projected onto a plane. If every particle in the target or projectile nucleus that is lined up with a part of the other nucleus subsequently emerges from the collision (i.e. all the *participants* and none of the *spectators*) the quantity $q(\alpha, \beta)$ can be written as

$$q(\alpha, \beta) = (\alpha + \beta)\, \varrho, \tag{11.12}$$

where ϱ is the nuclear particle number density. The total cross section for such particles is

$$\sigma_t = \pi \left(A_1 R_2^2 + A_2 R_1^2\right), \tag{11.13}$$

which is obtained by simply inserting (11.12) into (11.11). This cross section is much larger than the purely geometrical reaction cross section

$$\sigma_{\text{react}} = \pi(R_1 + R_2)^2, \tag{11.14}$$

because it counts the individual emerging participant particles. In fact, Eqs.

121

(11.13) and (11.14) can be combined to define the quantity $\langle nucleons \rangle_{\text{fireball}}$ which is the mean number of participant nucleons per collision averaged over impact parameter

$$\sigma_t = \langle nucleons \rangle_{\text{fireball}} \cdot \sigma_{\text{react}} \,. \tag{11.15}$$

11.2.2 Firestreak Model

If the fireball model concept is generalized in such a way that individual collinear rows of target and projectile particles combine to form mini-fireballs [Myers 78], then the velocity shear across the collision region and the diffuseness of the nuclear surface can be treated. One way of implementing such an approach is to rewrite (11.1) as

$$F_j(\boldsymbol{p}) = \int\limits_0^1 \mathrm{d}\eta \, Y(\eta) \, J_{\boldsymbol{p}' \to \boldsymbol{p}}[\beta_f(\eta)] \, f_j[\boldsymbol{p}'(\beta_i, \eta)] \,, \tag{11.16}$$

where the functional dependences of J and f on the quantity η are displayed,

$$\eta = \frac{\text{number of particles from the projectile}}{\text{number of projectile plus target particles}} \,, \tag{11.17}$$

and the *yield function* $Y(\eta)$ is defined by

$$Y(\eta) = \int \mathrm{d}\boldsymbol{s} \int \mathrm{d}\sigma \, w_b(x,y) \, \delta(\eta - \eta_b(x,y)) \,. \tag{11.18}$$

The quantity Y is presented here because of its pure geometrical character. It is the function describing the relative importance of different numbers of projectile particles on target particles (expressed in terms of η) after summing over the target and projectile overlap region in the x-y plane and summing over impact parameter.

Of course, the total cross section for a collision of one row of nucleons on another row of nucleons times the number of nucleons involved can be obtained from

$$\sigma_t = \int\limits_0^1 \mathrm{d}\eta \, Y(\eta) q(\eta) \,, \tag{11.19}$$

where $q(\eta)$ is a function which determines the fraction of the particles that participate for a given η. (For example, $q < 1$ when transparency effects are included).

11.2.3 Rows-on-Rows

In the rows-on-rows approach [Hüfner 77], [Knoll 79a] the general expression for the participant particle production cross section (11.1) is re-expressed as

$$E\frac{\mathrm{d}\sigma}{\mathrm{d}\boldsymbol{p}} = \sum_{MN} \sigma_{\text{PT}}(M,N) F_{MN}(\boldsymbol{p}) \,. \tag{11.20}$$

Here the geometrical weight factors $\sigma_{\text{PT}}(M,N)$ are the cross sections for finding a projectile row with M nucleons and a target row with N nucleons, respectively. The function $F_{MN}(\boldsymbol{p})$ is the one-nucleon momentum distribution resulting from the collinear collision of M projectile nucleons with N target nucleons. The cross sections $\sigma_{\mathrm{PT}}(M,N)$ are independent of the collision dynamics and only reflect the initial nuclear geometry and the size of the nucleon-nucleon scattering cross section $\sigma_{\mathrm{NN}}^{\mathrm{tot}}$. For inclusive cross sections (integrated over all impact parameters) they factorize into the respective nucleon-nucleus cross sections

$$\sigma_{\mathrm{PT}}(M,N) = \sigma_{\mathrm{P}}(M)\,\sigma_{\mathrm{T}}(N)/\sigma_{\mathrm{NN}}^{\mathrm{tot}}, \qquad (11.21)$$

where $\sigma_{\mathrm{A}}(N)$ is the incident nucleon cross section for encountering N nucleons in a collision with nucleus A. (Here A=P or T.) This cross section can be calculated by noting that the average number of nucleons encountered by an incident nucleon with impact parameter $\boldsymbol{s}$ is given by

$$\overline{N}_{\mathrm{A}}(\boldsymbol{s}) = \sigma_{\mathrm{NN}}^{\mathrm{tot}} \int \mathrm{d}z\, \varrho_{\mathrm{A}}(\boldsymbol{s},z), \qquad (11.22)$$

where $\varrho_{\mathrm{A}}(\boldsymbol{r})$ is the nuclear density distribution and the z-axis is along the beam. The cross section $\sigma_{\mathrm{A}}(N)$ is then obtained by integrating the corresponding Poisson distribution over all impact parameters,

$$\sigma_{\mathrm{A}}(N) = \int \mathrm{d}\boldsymbol{s}\, \frac{1}{N!}\left(\overline{N}_{\mathrm{A}}(\boldsymbol{s})\right)^{N} \exp\left(-\overline{N}_{\mathrm{A}}(\boldsymbol{s})\right). \qquad (11.23)$$

The total nucleon-nucleus reaction cross section which follows from summing (11.23) over all $N \geq 1$ can be written as

$$\begin{aligned}
\sigma_{\mathrm{A}} &= \sum_{N=1}^{\mathrm{A}} \sigma_{\mathrm{A}}(N) \\
&= \int \mathrm{d}\boldsymbol{s}\left[1 - \exp\left(-\overline{N}_{\mathrm{A}}(s)\right)\right]. \qquad (11.24)
\end{aligned}$$

The formulae (11.22–11.24) constitute the well known Eikonal expressions for the scattering of a hadron on a nucleus as derived by Glauber [Glauber 59]. This approach has been generalized to Eqs. (11.20) and (11.21) in [Hüfner 77] on the basis of a formalism developed by Glauber and Matthiae [Glauber 70]. Applications can be found in [Knoll 79a] and [Knoll 79b].

The integrated participant cross section can be given in closed form,

$$\sigma_t = \int \frac{\mathrm{d}\boldsymbol{p}}{E} \cdot E\frac{\mathrm{d}\sigma}{\mathrm{d}\boldsymbol{p}} = A_{\mathrm{P}}\sigma_{\mathrm{T}} + A_{\mathrm{T}}\sigma_{\mathrm{P}}, \qquad (11.25)$$

which is a generalization of (11.13). This result follows from Eqs. (11.20–11.24) if one observes that

$$\int \frac{\mathrm{d}\boldsymbol{p}}{E}\, F_{MN}(\boldsymbol{p}) = M + N \qquad (11.26)$$

and

$$\sum_{N=0}^{A} N\,\sigma_{\mathrm{A}}(N) = A\sigma_{\mathrm{NN}}^{\mathrm{tot}}. \qquad (11.27)$$

The form (11.25) has the advantage of including transparency effects through

(11.24) which will lead (especially for light nuclei) to smaller cross sections than the geometrical form (11.13). Unfortunately there is no simple closed form expression for the reaction cross section to improve the rough estimate (11.14) based on a knowledge of the density distributions.

An improvement of (11.23) which appears as Eq. (14) in [Schürmann 79] can be used to extend this approach to lower energies where the breakdown of the Eikonal approximation requires the inclusion of a correction for finite deflection angles in the scattering process.

11.2.4 Knock-on Collisions

In the firestreak or row-on-rows approach to high energy nuclear collisions of Sect. 11.2.3 an extreme assumption can be made [Hatch 79] that the nuclei are sharp surface spheres and that the individual nucleons scatter as free particles with the first other particle they encounter and that there are no secondary interactions. Then the total number of participants from a particular straight line trajectory is just twice the number of target (or projectile) particles (whichever is fewer). In this case the function q of (11.12) is given by

$$q_{\text{knock on}}(\alpha, \beta) = 2 \cdot \min(\alpha, \beta) \tag{11.28}$$

and the total cross section for the production of such particles is obtained by inserting this expression into (11.11) and performing the integration. The result is

$$\sigma_{\text{knock on}} = 2\pi A_1 \left(R_2^2 - \frac{1}{5} R_1^2 \right) , \tag{11.29}$$

where $R_2 \geq R_1$.

11.3 Density Distributions

When the approaches described in the previous section are extended beyond the assumption of uniform nuclear density distributions with a sharp surface, the standard diffuse surface distributions of Chap. 4 are usually employed. Historically the Fermi function (4.15) is the most common even though there are advantages associated with alternatives such as the symmetrized Fermi distribution (4.40) or the distributions generated by folding such as (4.65) and (4.69).

11.3.1 Density Projected on a Plane

When a diffuse nuclear density distribution is projected onto a plane (this is the relevant quantity for the models described earlier in Sect. 11.2), an excellent approximation for the distribution function is available in terms of a hemisphere smoothly connected to an exponential tail [Cecil 80].

The number of particles per unit area

$$\xi(s) = \varrho_0 \, \alpha(s) \tag{11.30}$$

in terms of the density of nuclear matter ϱ_0 and the radial coordinate s, can be written as

$$\alpha(s) = \begin{cases} 2\sqrt{R_s^2 - s^2} & , \ s \leq s' \\ c_1 \exp\left[-c_2(s - s')\right] & , \ s \geq s' \, . \end{cases} \tag{11.31}$$

If we set the range of the exponential to $c_2 = \sqrt{2}$ fm^{-1} (accurate to within a few percent) and then match the value and slope of this function at the transition point s' and demand the correct normalization (so that $A = 4\pi \varrho_0 R_0^3/3$), we find that

$$\begin{aligned} c_1 &= 2x \\ s' &= \sqrt{2}\, x^2 \\ R_s &= x^2 + 2x^4 \, , \end{aligned} \tag{11.32}$$

where x is obtained from the solution of the equation

$$R_0^3 = \left(x^2 + 2x^4\right)^{3/2} + \frac{3}{2}x + 2x^3 \tag{11.33}$$

and all lengths are expressed in fm.

Bibliography

[Adeev 71]   G.D. Adeev, I.A. Gamalya and P.A. Cherdantsev, Yad. Fiz. **14** (1971) 1144

[Aguilar 86]   M. Aguilar-Benitez, et al., Phys. Lett. **170B** (1986) 1

[Albrecht 73]   K. Albrecht, Nucl. Phys. **A207** (1973) 225

[Arseniev 68]   D.A. Arseniev, L.A. Malov, V.V. Pashkevich, A. Sobiczewski and V.G. Soloviev, Yad. Fiz. **8** (1968) 883

[Balazs 78]   E. Balazs and V.V. Pashkevich, Yad. Fiz. **27** (1978) 649

[Baltz 82]   A.J. Baltz and B.F. Baymann, Phys. Rev. **C26** (1982) 1969-83

[Barret 70]   R.C. Barret, Phys. Lett. **33B** (1970) 388-90

[Beringer 61]   R. Beringer and W.J. Knox, Phys. Rev. **121** (1961) 1195

[Beringer 63]   R. Beringer, Phys. Rev. **131** (1963) 1402-06

[Błocki 77]   J. Błocki, J. Randrup, W.J. Swiatecki and C.F. Tsang, Ann. Phys. (N.Y.) **105** (1977) 427-462

[Błocki 78]   J. Błocki, Y. Boneh, J.R. Nix, J. Randrup, M. Robel, A.J. Sierk and W.J. Swiatecki, Ann. Phys. **113** (1978) 330-386

[Błocki 81]   J. Błocki and W.J. Swiatecki, Ann. Phys. (N.Y.) **132** (1981) 53-65

[Błocki 82]   J. Błocki and W.J. Swiatecki, Berkeley Report LBL-12811 (1982)

[Bohr 39]   N. Bohr and J.A. Wheeler, Phys. Rev. **56]** (1939) 426-50

[Bohr 52]   A. Bohr, Dan. Mat. Fys. Medd. **26** (1952) 14

[Bohr 69]   A. Bohr and B.R. Mottelson, Nuclear Structure, Vol. I, Benjamin, Reading, 1969

[Bohr 75]   A. Bohr and B.R. Mottelson, Nuclear Structure Vol.II, Benjamin, Reading, 1975, p.47

[Bolsterli 72]   M. Bolsterli, E.O. Fiset, J.R. Nix and J.L. Norton, Phys. Rev. **C5** (1972) 1050

[Böning 87]   K. Böning, Z. Patyk, A. Sobiczewski, B. Nerlo-Pomorska and K. Pomorski, Acta Phys. Pol. **B18** (1987) 47-67

[Brack 72]   M. Brack, J. Damgaard, A.S. Jensen, H.C. Pauli, V.M. Strutinskii and C.Y. Wong, Rev. Mod. Phys. **44** (1972) 320-406

[Brack 74]   M. Brack, T. Ledergerber and H.C. Pauli, Nucl. Phys. **A234** (1974) 185

[Brack 85]   M. Brack, C. Guet and H.B. Håkansson, Phys. Rep. **123** (1985) 275-364

[Brosa 80]   U. Brosa and H.J. Krappe, Quart. J. Mech. Appl. Math. **33** (1980) 159-77

[Brown 80]   R.A. Brown and L.E. Scriven, Phys. Rev. Lett. **45** (1980) 180-183

[Budzanowski 76]   A. Budzanowski and A. Kapuścik eds., Radial Shape of Nuclei, Proc. 2nd Nucl. Phys. Div. Conf. of the European Physical Society, Cracow, Poland June 1976, Jagellonian Univ. Inst. Nucl. Phys., Cracow, 1976

[Businaro 55a]   U.L. Businaro and S. Gallone, Nuovo Cim. **1** (1955) 629

[Businaro 55b]   U.L. Businaro and S. Gallone, Nuovo Cim. **1** (1955) 1277

[Businaro 57]   U.L. Businaro and S. Gallone, Nuovo Cim. **5** (1957) 315

[Carlson 61a]   B.C Carlson, Iowa State J. Sci. **35** (1961) 319-330

[Carlson 61b]   B.C Carlson, J. Math. Phys. **2** (1961) 441

[Carlson 63]   B.C Carlson and G.L. Morley, Am. J. Phys. **31** (1963) 209-211

[Cecil 80]	G. Cecil, S. Das Gupta and W.D. Myers, Phys. Rev. **C22** (1980) 2018-23
[Chandrasekhar 59]	S. Chandrasekhar, Proc. London Math. Soc. 9 (1959) 141
[Chandrasekhar 61]	S. Chandrasekhar, The Oscillations of a Viscous Liquid Globe, Clarendon, Oxford, 1961
[Chasman 70]	R.C. Chasman, Phys. Rev. **A1** (1970) 2144
[Cohen 62]	S. Cohen and W.J. Swiatecki, Ann. Phys. (N.Y.) **19** (1962) 67-164
[Cohen 63]	S. Cohen and W.J. Swiatecki, Ann. Phys. (N.Y.) **22** (1963) 406-37
[Cohen 74]	S. Cohen, F. Plasil and W.J. Swiatecki, Ann. Phys. **82** (1974) 557
[Collard 67]	H.R. Collard, L.R.B. Elton and R. Hofstadter, Nuclear Radii, Vol. 2, Group I Landolt Börnstein, Numerical Data and Functional Relationships in Science and Technology, Springer, Berlin, 1967
[Damgaard 69]	J. Damgaard, H.C. Pauli, V.V. Pashkevich and V.M. Strutinsky, Nucl. Phys. **A135** (1969) 432
[Davies 75]	K.T.R. Davies and A.J. Sierk, J. Comp. Phys. **18** (1975) 311
[Davies 76a]	K.T.R. Davies and J.R. Nix, Phys. Rev. **C14** (1976) 1977-94
[Davies 76b]	K.T.R. Davies, A.J. Sierk and J.R. Nix, Phys. Rev. **C13** (1976) 2385
[Devries 75]	R.M. Devries and M.R. Clover, Nucl. Phys. **A243** (1975) 528-32
[Diehl 74]	H. Diehl and W. Greiner, Nucl. Phys. **A229** (1974) 29
[Eisenberg 70]	J.M. Eisenberg and W. Greiner, Nuclear Models, North Holland, Amsterdam, 1970
[Elton 61]	L.R.B. Elton, Nuclear Sizes, Oxford Univ. Press, Oxford, 1961
[Feldmeier 80]	H. Feldmeier, Proc. of the XIVth School in Nuclear Physics, Mikołajki 1979, Part 2, Nukleonika **25** (1980) 171-75
[Foland 59]	W.D. Foland and R.D. Present, Phys. Rev. **113** (1959) 613
[Ford 54]	K.W. Ford and D.L. Hill, Phys. Rev. **94** (1954) 1630-37
[Ford 69]	K.W. Ford and J.G. Wills, Phys. Rev. **185** (1969) 1429-38
[Ford 73]	K.W. Ford and G.A. Rinker, Jr., Phys. Rev. **C7** (1973) 1206-21
[Friedrich 72]	J. Friedrich and F. Lenz, Nucl. Phys. **A183** (1972) 523-44
[Friedrich 82]	J. Friedrich and N. Voegler, Nucl. Phys. **A373** (1982) 192-224
[Gaudin 74]	M. Gaudin, J. Physique **35** (1974) 885
[Geilikman 55]	B.T. Geilikman, Proc. UN Int. Conf. on the Peaceful Uses of Atomic Energy, Vol. 2, p.201
[Geilikman 58]	B.T. Geilikman, Proc. 2nd. UN Int. Conf. on the Peaceful Uses of Atomic Energy, Geneva 1958, New York, 1958, Vol. 15, paper P/2474, p.273
[Geilikman 58a]	B.T. Geilikman, Proc. 2nd. UN Int. Conf. on the Peaceful Uses of Atomic Energy, Geneva 1958, New York 1958, Vol. 15, paper P/2473, p.279
[Glauber 59]	R.J. Glauber, Lectures in Theoretical Physics (W.E. Brittin et al. eds.), Vol. I, p.315, Interscience, New York, 1959
[Glauber 70]	R.J. Glauber and G. Matthiae, Nucl. Phys. **B21** (1970) 135
[Gogny 77]	D. Gogny and R. Padjen, Nucl. Phys. **A293** (1977) 365
[Gosset 77]	J. Gosset, H.H. Gutbrod, W.G. Meyer, A.M. Poskanzer, A. Sandoval, R.Stock and G.D. Westfall, Phys. Rev. **C16** (1977) 629-57
[Gosset 78]	J. Gosset, J.I. Kapusta and G.D. Westfall, Phys. Rev. **C18** (1978) 844-55
[Götz 72]	U. Götz, H.C. Pauli, K. Alder and K. Junker, Nucl. Phys. **A192** (1972) 1
[Grammaticos 73]	B. Grammaticos, Phys. Lett. **443** (1973) 343
[Grammaticos 82]	B. Grammaticos, Z. Phys. **A305** (1982) 257-62
[Gray 19]	A. Gray, Phil. Mag. Ser. 6 **38** (1919) 201
[Griffin 86]	J.J. Griffin and M. Dworzecka, Nucl. Phys. **A455** (1986) 61
[Guet 80]	C. Guet, R. Bengtsson and M. Brack, in [P C F 80] Vol. II p. 411-22 Phys. and Chem. of Fission 1979, Vol. II, IAEA, Vienna, 1980, Paper IAEA-SM/24-H3

[Gunter 59]	W.D. Gunter, Jr. and R.A. Hubbs, Phys. Rev. **113** (1959) 252
[Hasse 68a]	R.W. Hasse, R. Ebert and G. Süssmann, Nucl. Phys. **A106** (1968) 117-128
[Hasse 68b]	R.W. Hasse, Nucl. Phys. **A118** (1968) 577-591
[Hasse 69]	R.W. Hasse, Nucl. Phys. **A128** (1969) 609
[Hasse 71]	R.W. Hasse, Ann. Phys. (N.Y.) **68** (1971) 377-461
[Hasse 75]	R.W. Hasse, Ann. Phys. (N.Y.) **93** (1975) 68
[Hasse 77]	R.W. Hasse, Habilitationsschrift, Univ. München (1977)
[Hasse 78]	R.W. Hasse, Pramāṇa (India) **11** (1978) 441-55
[Hatch 79]	R.L. Hatch and S.E. Koonin, Phys. Lett. **81B** (1979) 1-4
[Helm 56]	R.H. Helm, Phys. Rev. **104** (1956) 1466-75
[Hill 53]	D.L. Hill and J.A. Wheeler, Phys. Rev. **89** (1953) 1102-45
[Hill 54]	D.L. Hill and K.W. Ford, Phys. Rev. **94** (1954) 1617-29
[Hirschfelder 54]	J.O. Hirschfelder, C.F. Curtiss and R.B. Bird, Molecular Theory of Gases and Liquids, Chapter 12, pp. 846,906, Wiley, New York 1954
[Hüfner 77]	J. Hüfner and J. Knoll, Nucl. Phys. **A290** (1977) 460-92
[Jackson 75]	J.D. Jackson, Classical Electrodynamics, Wiley, New York 1962 (second ed. 1975)
[Jänecke 72a]	J. Jänecke, Nucl. Phys. **A181** (1972) 49-75
[Jänecke 72b]	J. Jänecke, in [Sanders 72] p. 221-35
[Kaniowska 76]	T. Kaniowska, A Sobiczewski, K. Pomorski and S.G. Rohozinski Nucl. Phys. **A274** (1976) 151
[Kelson 64]	I. Kelson, Phys. Rev. **136** (1964) B1667
[Kim 76]	Y.N. Kim, S. Wald and A. Ray, in [Budzanowski 76], Proc. of the Second Nuclear Physics Divisional Conference of the European Physical Society, Cracow, June 1976, p. 34-52
[Klepper 84]	O. Klepper, ed., Proc. 7th Int. Conf. on Atomic Masses and Fundamental Constants, Darmstadt-Seeheim, Technische Hochschule, Darmstadt, 1984
[Knoll 79a]	J. Knoll and J. Randrup, Nucl. Phys. **A324** (1979) 445-63
[Knoll 79b]	J. Knoll, Phys. Rev. **C20** (1979) 773-80
[Krappe 73]	H.J. Krappe and J.R. Nix, Proc. 3rd IAEA Symp. on the Physics and Chemistry of Fission, Rochester 1973, Vol. I, IAEA, Vienna, 974, p.159-75
[Krappe 76]	H.J. Krappe, Ann. Phys. (NY) **99** (1976) 142-63
[Krappe 79]	H.J. Krappe, J.R. Nix and A.J. Sierk, Phys. Rev. **C20** (1979) 992-1013
[Krappe 81]	H.J. Krappe, Proc. XIVth Mikołajki School of Nuclear Physics, Aug.30-Sept.12, 1981, B. Sikora and Z. Wilhelmi eds., Harwood Academic Publishers
[Krivine 81]	H. Krivine and J. Treiner, J. Math. Phys. **22** (1981) 2484-5
[Larsson 73]	S.A. Larsson, Physica Scripta **8** (1973) 17-31
[Lawrence 65]	J.N.P. Lawrence, Phys. Rev. **139** (1965) B1227
[Lawrence 67]	J.N.P. Lawrence, L A S L Report LA-3774(1967)
[Leander 74]	G. Leander, Nucl. Phys. **A219** (1974) 245
[Lindner 68]	A. Lindner, Z. Phys. **211** (1968) 195-212
[Lindner 69]	A. Lindner, Z. Phys. **219** (1969) 1-4
[Löbner 70]	K.E.G. Löbner, M. Vetter and V. Hönig, Nuclear Data Tables **A7** (1970) 495-564
[Martinot 77]	M. Martinot and M. Gaudin, Rev. Roum. Phys. **22** (1977) 17
[Möller 81]	P. Möller and J.R. Nix, Nucl. Phys. **A361** (1981) 117-46
[Moon 61]	P. Moon and D. Eberle Spencer, Field Theory Handbook, Springer, Heidelberg, 1961
[Münchow 79]	L. Münchow and H. Schulz, J. Phys. **G5** (1979) 527-40

[Myers 66]	W.D. Myers and W.J. Swiatecki, Nucl. Phys. **81** (1966) 1-60
[Myers 69]	W.D. Myers and W.J. Swiatecki, Ann. Phys. (N.Y.) **55** (1969) 395-505
[Myers 70]	W.D. Myers, Nucl. Phys. **A145** (1970) 387-400
[Myers 73]	W.D. Myers, Nucl. Phys. **A204** (1973) 465
[Myers 74]	W.D. Myers and W.J. Swiatecki, Ann. Phys. (N.Y.) **84** (1974) 186-210
[Myers 76]	W.D. Myers, Nucleonica **21** (1976) 3-28
[Myers 77]	W.D. Myers, Droplet Model of Atomic Nuclei, IFI / Plenum Data Co., New York, 1977
[Myers 78]	W.D. Myers, Nucl. Phys. **A296** (1978) 177-88
[Myers 82]	W.D. Myers and W.J. Swiatecki, Ann. Rev. Nucl. Part. Sci. **32** (1982) 309-34
[Myers 83]	W.D. Myers and K.-H. Schmidt, Nucl. Phys. **A410** (1983) 61-73
[Nilsson 55]	S.G. Nilsson, Dan. Mat. Fys. Medd. **29** (1955) 16
[Nilsson 69]	S.G. Nilsson, C.F. Tsang, A. Sobiczewski, Z. Szymański, S. Wycech, C. Gustafson, I.-L. Lamm, P. Möller and B. Nilsson, Nucl. Phys. **A131** (1969) 1-66
[Nix 64]	J.R. Nix, U C R L - 11338 (1964)
[Nix 65]	J.R. Nix and W.J. Swiatecki, Nucl. Phys. **71** (1965) 1-94
[Nix 67]	J.R. Nix, Ann. Phys. **41** (1967) 52,U C R L - 16786 (1966)
[Nix 68]	J.R. Nix, Berkeley Report U C R L - 17958 (1968)
[Nix 69]	J.R. Nix, Nucl. Phys. **A130** (1969) 241-92
[Nix 72]	J.R. Nix, Ann. Rev. Nucl. Sc. **22** (1972) 65
[Nossoff 55]	V.G. Nossoff, Proc. Int. Conf. on the Peaceful Uses of Atomic Energy, Geneva, 1955, vol.2(1955) p.205, United Nations, New York, 1956
[Pashkevich 71]	V.V. Pashkevich, Nucl. Phys. **A169** (1971) 275
[Pashkevich 83]	V.V. Pashkevich, Proc. Int. School-Seminar on Heavy Ion Physics, Alushta, April 1983 (JINR, Dubna 1983) p. 405
[Pauli 73]	H.C. Pauli, Phys. Rep. **7**, no. 2 (1973) 35-100
[P C F 80]	(no editor), Physics and Chemistry of Fission 4, Jülich, 1979, IAEA, Vienna, 1980
[Pik-Pichak 80]	G.A. Pik-Pichak, Yad. Fiz. **31** (1980) 98-108
[Present 40]	R.D. Present and J.K. Knipp, Phys. Rev. **57** (1940) 751 and 1188
[Present 46]	R.D. Present, F. Reines and J.K. Knipp, Phys. Rev. **70** (1946) 557
[Preston 62]	M.A. Preston, Physics of the Nucleus, Addison-Wesley, Reading Mass., 1962
[Quentin 69]	P. Quentin, J. de Physique **30** (1969) 497
[Remaud 78]	B. Remaud, Univ. Nantes Report LSNN-78-04 (unpublished) 1978
[Remaud 81]	B. Remaud and G. Royer, J. Phys. **A14** (1981) 2897-2910
[Ring 80]	P. Ring and P. Schuck, The Nuclear Many-Body Problem, Springer, Heidelberg, 1980
[Robinson 87]	R.L. Robinson, Science **235** (1987) 633
[Rohozinski 81]	S.G. Rohozinski and A. Sobiczewski, Acta Phys. Pol. **B12** (1981) 1001-7
[Royer 82]	G. Royer and B. Remaud, J. Phys. **G8** (1982) L159
[Royer 84]	G. Royer and B. Remaud, J. Phys. **G10** (1984) 1047
[Ryce 65]	S.A. Ryce and P.A. Patriarche, Can. J. Phys. **43** (1965) 2192
[Ryce 72]	S.A. Ryce, R.R. Wyman and A.T. Stewart Can. J. Phys. **50** (1972) 2217
[Sanders 72]	J.H. Sanders and A.H. Wapstra, editors, Atomic Masses and Fundamental Constants 4, Plenum, London, 1972
[Satchler 72]	G.R. Satchler, J. Math. Phys. **13** (1972) 1118-9
[Saupe 87]	G. Saupe, priv. comm. (1987)

[Schirmer 73]	J. Schirmer, S. Knaak and G. Süssmann, Nucl. Phys. **A199** (1973) 31
[Schultheis 75]	H. Schultheis and R. Schultheis, J. Math. Phys. **6** (1975) 905-909
[Schürmann 79]	B. Schürmann, Phys. Rev. **C20** (1979) 1607-11
[Schütte 75]	G. Schütte and L. Wilets, Nucl. Phys. **A252** (1975) 21
[Seeger 75]	P.A. Seeger and W.M. Howard, Nucl. Phys. **A238** (1975) 491-532
[Seyler 61]	R.G. Seyler and C.H. Blanchard, Phys. Rev. **124** (1961) 227 Mikołajki 1979, Part 2, Nukleonika **25** (1980) 171-75
[Sierk 80]	A.J. Sierk and J. R. Nix, Phys. Rev. **C21** (1980) 982-987
[Slater 60]	J.C. Slater, Quantum Theory of Atomic Structure, Vol. II, Chapter 22, McGraw Hill, New York, 1960
[Sobiczewski 69]	A. Sobiczewski, Z. Szymanski, S. Wycech, S.G. Nilsson, J.R. Nix, C.F. Tsang, C. Gustafson, P. Möller and B. Nilsson, Nucl. Phys. **A131** (1969) 67-91
[Srivastava 82a]	D.K. Srivastava, Phys Lett. **113B** (1982) 353-6
[Srivastava 82b]	D.K. Srivastava, Phys Lett. **112B** (1982) 289-91
[Srivastava 83]	D.K. Srivastava, Phys Lett. **122B** (1983) 18
[Stavinsky 68]	V.S. Stavinsky, N.S. Rabotnov and A.A. Seregin, Yad. Fiz. **7** (1968) 1051 [Sov. J. Nucl. Phys. **7** (1968) 631]
[Strutinsky 62]	V.M. Strutinsky, JETP (USSR) **42** (1962) 1571-81; [Sov. Phys. JETP **15** (1962) 1091-97]
[Strutinsky 63]	V.M. Strutinsky, N.Ya. Lyashenko and N.A. Popov, Nucl. Phys. **46** (1963) 639-59
[Strutinsky 68]	V.M. Strutinsky, Nucl. Phys. **A122** (1968) 1-33
[Süssmann 75]	G. Süssmann, Z. Phys. **A274** (1975) 145-59
[Swiatecki 56a]	W.J. Swiatecki, Phys. Rev. **101** (1956) 651-4
[Swiatecki 56b]	W.J. Swiatecki, Phys. Rev. **104** (1956) 993-1005
[Swiatecki 58]	W.J. Swiatecki, Paper No. P/651, Proc. Second Int. Conf. Peaceful Uses Atomic Energy, Geneva, 1958, Vol. 15 p. 248-72
[Swiatecki 80]	W.J. Swiatecki, Prog. Part. Nucl. Phys. **4** (1980) 383
[Swiatecki 81]	W.J. Swiatecki, Proc. Nobel Symp. on High Spin States, Arenas, Sweden, 1980, Physica Scripta **24** (1981) 113
[Treiner 86]	J.Treiner and H. Krivine Ann. Phys. (NY) **170** (1986) 406-53
[Trentalange 80]	S. Trentalange, S.E. Koonin and A.J. Sierk, Phys. Rev. **C22** (1980) 1159-67
[Viñas 75]	F.J. Viñas and G. Madurga, Nucl. Phys. **A240** (1975) 109-19
[Weizsäcker 35]	C.F. von Weizsäcker, Z. Physik **96** (1935) 431
[Wilets 64]	L. Wilets, Theories of Nuclear Fission, Clarendon Press, Oxford, 1964
[Wilkins 76]	B.D. Wilkins, E.P. Steinberg and R.R. Chasman, Phys. Rev. **C14** (1976) 1832-63
[Wong 73]	C.Y. Wong, Ann. Phys. (N.Y.) **77** (1973) 279-353
[Wong 78]	Y. Wong, Phys. Rev. **C17** (1978) 331-40

Citation Index

Subject Index